Aerospace Transmission Systems

Seminar Organizing Committee

Organizing Committee Member

J Forfar CEng, MIMechE
Rolls Royce Aerospace Group
Derby, UK

Transmission Technical Activity Committee Chairman

P Morris CEng, MIMechE
Consultant Engineer
Gloucester, UK

Papers presented at a one-day seminar *Concurrent Design and Manufacture of Aerospace Transmission Systems* on 20 May 1998 held at Rolls-Royce, Derby, UK

IMechE
Seminar Publication

I MECH E

Aerospace Transmission Systems
– Concurrent Design and Manufacture

20 May 1998
Rolls-Royce, Derby, UK

Organized by the Aerospace Industries Division of
the Institution of Mechanical Engineers (IMechE)

IMechE Seminar Publication 1998–6

**Professional
Engineering
Publishing**

Published by Professional Engineering Publishing Limited for
The Institution of Mechanical Engineers, Bury St Edmunds and London, UK.

First Published 1998

ISSN 1357–9193
ISBN 1 86058 161 7

A CIP catalogue record for this book is available from the British Library.

Printed by Antony Rowe Limited, Chippenham, Wilts, UK.

Contents

Related Titles of Interest

Title	Editor/Author	ISBN
Gas Turbines – Reducing Time and Cost from Concept to Product	IMechE Seminar 1997–13	1 86058 126 9
The Manufacturing Challenge in Aerospace	IMechE Seminar 1997–3	1 86058 111 0
Avionic Systems, Design, and Software (Aerotech '95)	IMechE Seminar 1996–11	1 86058 045 9
Aircraft Structures and Materials (Aerotech '95)	IMechE Seminar 1996–10	1 86058 044 0
Aircraft Health and Usage Monitoring Systems (Aerotech '95)	IMechE Seminar 1996–9	1 86058 043 2
Multi-Body Dynamics: Monitoring and Simulation Techniques	H Rahnejat and R Whalley	1 86058 064 5

For the full range of titles published by Professional Engineering Publishing contact:

Sales Department
Professional Engineering Publishing Limited
Northgate Avenue
Bury St Edmunds
Suffolk
IP32 6BW
UK

Tel: 01284 724384
Fax: 01284 718692

S520/001/98

Functional simulation and design optimization of the integrated step aside gearbox and intercase concept

C SMITH, M GREENHALGH, and I BOSTON
Parametric Technology Limited, Gateshead, UK

This paper describes the use of the Pro/MECHANICA Design Optimisation System during the early stages of the design of a new integrated Step Aside Gearbox (SAGB) and Intercase concept for a new large aero gas turbine.

The system was used to automatically evaluate the sensitivity of the design, in terms of stiffness and stress levels, to various changes in geometry. This identified the critical design features that were then used to optimise the concept. This optimisation was successful in reducing the weight of the design concept by 15% whilst ensuring the component remained serviceable.

This investigation illustrates how design optimisation may be used, within a Concurrent Engineering Environment to improve product quality, reduce development lead times and costs.

1. Introduction

1.1. Project Background

The objective of this project was the redesign of the Step Aside Gearbox for a new, large aero gas turbine. The existing design comprised of a bolt on cast component, which is shown in Figure 1 below.

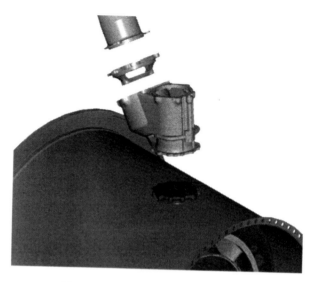

Figure 1 - Existing SAGB Design

Experience with the existing gearbox design had highlighted a sensitivity of performance to the stiffness of the assembly. It was thought this was due to the compliance of the structure, resulting from the small footprint at the interface with the intermediate casing.

The integration of the Step Aside Gearbox with the Intermediate Casing casting was identified as a promising design enhancement. It was proposed that the removal of the bolted interface would significantly increase the stiffness of the assembly.

As the intermediate casing has one of the longest manufacturing lead times of any component in the engine, any delay in its definition has serious implications for the entire program. Due to these time pressures, a traditional design-analyse-redesign loop (Figure 2) cannot be adopted, and often several costly prototypes are required. One solution to this problem is to conduct functional simulation and optimisation concurrently with concept design, as shown in Figure 3. In this case study Pro/MECHANICA was used to facilitate this process.

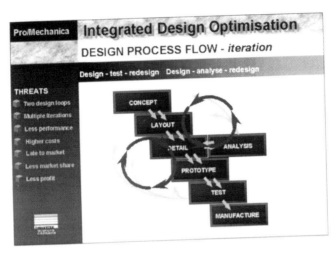

Figure 2 - Traditional Design - Analyse - Redesign Process

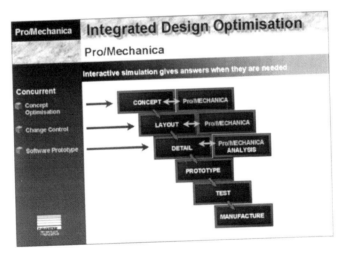

Figure 3 - Concurrent Functional Simulation and Design Optimisation

1.2. Pro/MECHANICA Overview

Pro/MECHANICA is a system for the functional simulation and optimisation of designs. The system comprises of three main components, the Pro/ENGINEER Parametric, Feature Based solid modeller, an analysis engine based upon Geometric Element Analysis (GEA) and algorithms for the calculation of design sensitivities and optimisation.

1.2.1. Pro/ENGINEER

The Pro/ENGINEER solid modeller facilitates the rapid generation of design geometry that responds predictably and robustly to change. This is important in the early stages of the design process where there is little time available for geometry modelling and the design itself is very fluid. Furthermore, the process of automatic design optimisation requires that the analysis system itself modifies the CAD model, without any intervention from the designer, and thus a robust geometric definition of the geometry is required.

1.2.2. Geometric Element Analysis

The analysis "engine" of the software is based upon a refinement of the Finite Element Method termed Geometric Element Analysis (GEA). This method allows the mathematical descriptions of the finite elements to be adapted automatically during the solution process in order to achieve a solution, converged to a required accuracy. Additionally, Geometric Elements may represent more complex geometry than traditional finite elements and are far less sensitive to distortions from their ideal shape. The Geometric Elements may be created automatically using the AutoGEM meshing facility.

These inherent properties of GEA allow the non-specialist to create a valid analysis model. Thus, utilising this technique, the designer is empowered to conduct functional simulation early in the design process. The consistent accuracy of the GEA method is also vital in achieving convergence in Design Optimisation studies.

1.2.3. Sensitivity and Optimisation Studies

The parametric nature of the model and the automatic Geometric Element Analysis may be exploited to investigate the consequence of possible design changes. The designer may select critical design dimensions and input their upper and lower limits. Pro/MECHANICA may then calculate design curves of how a measure, such as peak stress, varies through this range. This Sensitivity Analysis aids understanding of the important features of the design concept and helps identify potential areas for improvement.

The system may also be used to perform automatic Design Optimisation studies. In this case the Designer selects the critical design dimensions (Design Variables) and ranges, specifies a goal; such as minimise mass, and design constraints, such as maximum stress in the model. Pro/MECHANICA will then automatically iterate the design to find the optimum solution.

2. Preliminary Design Definition

The first stage of the design study concerned the creation of a parametric, three-dimensional representation of the design concept (Figure 4). The modeller was used to quickly create a $72°$ segment of the intercase beginning at the outer swan neck duct and extending to the outercase. The lower housing of the SAGB was then defined and merged onto the intercase to form a single part. The upper and lower bearing mounting flanges were then incorporated into the SAGB model. These were the main components to be optimised.

Figure 4 - Concept Model

The geometry was generated in a simplified form to facilitate rapid analysis. The main simplification was to enforce constant wall thicknesses in the model as the solid geometry was to be represented as a shell for simulation purposes.

The initial model definition is shown in Figure 5 and Figure 6 It may be clearly seen how the bearing flanges have been incorporated into the intercase as a single component. The simple flange definition was considered sufficient for a true evaluation of the load path associated with the bearing mounting forces. It was assumed that any production geometry, such as blends and chamfers, would add to the rigidity of the assembly.

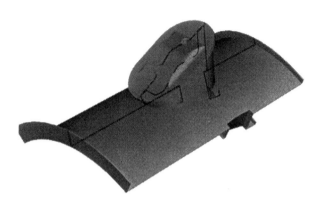

Figure 5 - Merged Model

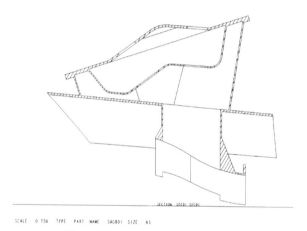

SCALE 0 750 TYPE PART NAME SAG801 SIZE A3

Figure 6 - Section through Merged Model

3. Initial Stress Analysis

3.1. Modelling Assumptions

For simulation purposes, the model was to be represented as a shell. The major advantage of this approach was reduction in meshing and analysis times. Such models often contain up to 100+ loadcases and Sensitivity and Optimisation studies often involve many iterations and associated analysis runs, thus it is prudent to keep models as simple as possible initially. The shell analysis required the definition of the mid-surface of the solid model. The modeller allows the definition of this surface by selecting pairs of surfaces on the outer and inner faces of the solid. The shell elements are created on this mid-plane as part of the analysis process.

3.2. Boundary Conditions

The model was constrained at the base of the swan neck duct to represent the remainder of the relatively stiff intercase. This assumption had been used previously in similar analyses of this type previously. Symmetry constraints were applied and the flanges at either end of the model were fixed in the axial direction. Thermal and bearing loads were applied to the model.

Throughout the study, the modelling assumptions made were discussed with stress analysts within the company.

3.3. Analysis Runs

The model was then run through a series of bearing load cases in order to establish the worst case combination to be used in the sensitivity and optimisation studies. The run time for each load combination was in the order of ten minutes. These preliminary results demonstrated that the deflection due to the thermal load dominated. It was therefore decided to remove the thermal load in order to isolate the effects of the bearing loads. Typical displacement and stress plots are shown in Figure 7 and Figure 8 below. These results confirmed that the stresses in the bearing flanges were not significant and the displacements were of the prime concern.

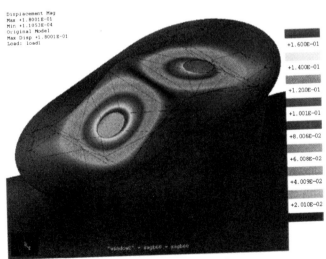

Figure 7 - Displacement Results

S520/001

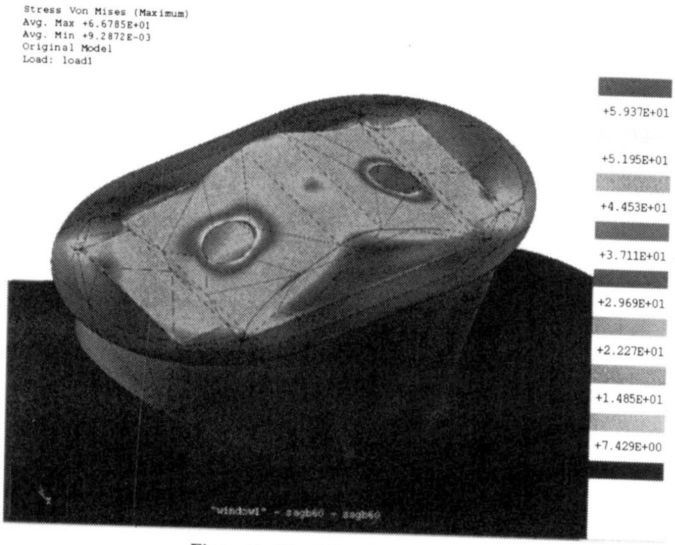

```
Stress Von Mises (Maximum)
Avg. Max +6.6785E+01
Avg. Min +9.2872E-03
Original Model
Load: load1
```

+5.937E+01

+5.195E+01

+4.453E+01

+3.711E+01

+2.969E+01

+2.227E+01

+1.485E+01

+7.429E+00

Figure 8 - Von Mises Stress Results

4. Design Sensitivity Study

The bearing support and bearing mounting flanges were identified as key components in the design. In the initial design definition the thickness of these components was arbitrarily chosen. It was thought that these components might contribute significantly to the mass of the assembly and thus a sensitivity study was conducted to establish the consequences of changing these design variables. The top flange of the SAGB was also included in the study, however the remainder of the component was constrained by manufacturing limitations. The sensitivity curves calculated for this study are shown in Figure 9 below. It was apparent from these results that the thickness of the bearing support flange had the greatest influence on the performance of the design. Similarly it could be seen that the thickness of the top flange was relatively insignificant.

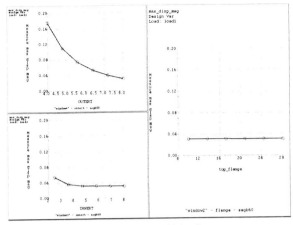

Figure 9 - Design Sensitivity Curves

5. Optimisation Study

A Design Optimisation study was defined using the critical force combination established from the initial stress analysis and design variable information from the sensitivity study. It had been shown that the maximum stress was well within permitted levels and thus the design constraint was the displacement of the bearing support structures. If the deformation of the mounting flanges were too large this would lead to bearing vibration problems and eventually shaft whirl.

Optimisation studies were completed for Aluminum and Titanium concepts. The maximum deflection was specified as a design constraint and reduction in mass as the optimisation goal. The results from these studies are summarised in Figure 10 below.

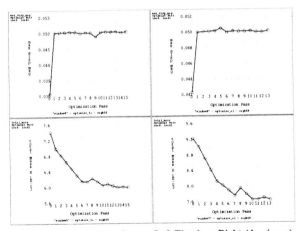

Figure 10 - Optimisation Curves (Left Titanium, Right Aluminum)

Both studies yielded weight savings in the region of 15%, whilst keeping the deflections within acceptable limits.

6. Conclusion

This study demonstrates that functional simulation and optimisation of designs at the concept stage by Design Engineers is achievable. In this case Pro/MECHANICA has been successfully applied to facilitate this process.

The concept design was developed from a two dimensional scheme, validated and optimised within a two week period. This study was completed while the project was in the concept stage and thus the changes could be accommodated before commitment was made to costly downstream activities such as tooling, detailed design and physical testing.

The guidance given by the stress analysts within the company was extremely important. Whilst the system facilitates simulation and optimisation of designs by Design Engineers the guidance of analysts should always be sought to ensure that valid assumptions have been made.

The three key technologies facilitating early insight and improvement of this design were identified as: -

- Rapid and robust parametric modelling facilitated by the solid modeller
- Ease of analysis model creation facilitated by the geometry abstraction tools in the modeller and the GEA analysis technology.
- Design improvement driven by Sensitivity and Optimisation studies.

The optimisation was successful in reducing the weight of the concept design by 15% whilst ensuring the component remained serviceable.

© With Author

Key system approach to the concurrent definition, verification, and manufacture of main shafts for gas turbine aeroengines

J W TAYLOR and **A C TALBOT**
Aerospace Group, Rolls-Royce plc, UK

1.0 INTRODUCTION

A Key System is defined as an integration of various technical systems involved in engineering and manufacture at an engine or component level. Such a system may include CAD/CAM, analytical, data access and knowledge based systems. This paper focuses at a component level on main shafts for gas turbine aero engines and describes a Key System which guides the sequence of activities, incorporates best practice and improves data communication to provide a seamless process based on the concept of a single electronic model through the component development life cycle from initial concept to final manufacture. The objective is to achieve a shorter time to market, a right first time end product and reduced operating costs with the same level of resource in a more process-focused manner by greater use of concurrent engineering.

2.0 THE NEED FOR CONCURRENT ENGINEERING OF ENGINE MAIN SHAFTS

The prevailing market forces result in a continuing need for engines with improved efficiencies and reduced cost and weight generating a trend for higher bypass ratios by reducing engine core size in either relative or absolute terms. Consequently there is an ever greater duty required of main shafts due to higher temperature and higher shaft speed engine cores and significantly increased shaft torques in the LP spool. A typical large Rolls-Royce engine is configured with three concentric spools where the LP shaft system is at the smallest diameter. The primary function of the shaft system is to transmit torque to allow the turbines to drive the respective compressors. The LP turbine shaft also carries a central oil tube that allows delivery of oil to the engine thrust bearings via radial holes in the shaft. The annular cavity between the concentric LP and IP turbine shafts acts as an air duct to allow the supply of air for sealing and cooling to be distributed through radial holes in the IP turbine shaft. Shafts are subjected to a load set that is generally dominated by torque but also includes axial loads due to thrust, and bending loads due to gyroscopic effects and engine carcass deflections during normal flight conditions. The load set can have different proportions for occasional more severe flight conditions which need to be designed for to ensure fitness for purpose and to meet the requirements of the certifying authorities.

To define a shaft system for a new engine project encompasses many and varied activities including material selection, strength, fatigue, thermal and dynamics analyses for a variety of load cases, air system and oil system requirements, manufacturability and the generation of the many electronic models and definitions needed to satisfy the analysis and detail activities. Shorter time to market can result in being first on the wing for aircraft certification and therefore gaining a significant reduction in the cost of flight certification testing, as well as certifying ahead of competitors.

Clearly there is an opportunity for a great deal of concurrent engineering to shorten the overall time scale, to optimise the shaft definition before a premature space constraint is created by other components and to ensure a right first time end product by improving the communication of data. This can be achieved successfully with a team of experienced and knowledgeable engineers using a variety of tools and data sources (reference 5.1). The Key System approach will increase the level of success over a traditional process using a similar level of resource by integrating the various technologies involved to enable interfacing at all levels of the concurrent engineering process and by creating a system which ensures every part of the process accesses a common electronic definition of the component.

3.0 SHAFT KEY SYSTEM

3.1 Description

The Key System approach is to create a computer based process by integrating packages that were previously stand alone with a centrally located single electronic definition of a component, in this case a shaft. This single electronic definition is conceptually surrounded by the various technologies and computer software packages that have read access to create the appropriate model for their process. The Shaft Key System is represented in simplified form by figure 1. The system will exist in a common environment deployed to standard desktop workstations connected to the company network and access to the various software packages will be via an Intranet using Internet Technology. Some of the packages exist on the PC network and some information databases exist on the mainframe computing system, and these are integrated with the Key System by links between the networks. The components of the system will include the equivalent of methods documents in electronic form, planning and resource tools, experience databases to assist methods application, finite element analysis software packages, CAD modelling software and component specific standard networks.

The Shaft Key System differs from an Artificial Neural Network approach (reference 5.2) in that many decisions are made by manual intervention , and it is the integration of improved processes in a user-friendly environment that derives the benefits to the business by encouraging a culture of concurrent engineering.

3.2 Standard Network

The Standard Network is a statement of current capability and provides the basis for programme management. It is a template of the processes and tasks involved throughout the definition , verification and manufacture of a component. The network model uses concepts that have existed for many years in the form of PERT (Programme Evaluation and Review Technique) programming and Critical Path Analysis to identify a typical task list for a main shaft and demonstrate the interdependencies of these tasks. The computer model of the Standard Network also contains a data sheet for each task to record sources of technical data required and the likely duration and resource level to perform the task.

The Standard Network model promotes concurrent engineering by defining the horizontal relationships between organisational functions and the vertical relationships by integrating the activities performed within the functions. This can be used for clean sheet designs or to evaluate alternative improvement proposals as a sub-set of the total process by selecting appropriate groups of activities.

3.3 Capability Database

The Capability Database is an Intranet framework populated with the technical and procedural processes relevant to the tasks contained in the Standard Network, which the user can call up at the time and point of need. The database is a computerised version of traditional paper methods documents with on-screen calculation aids such as live formulae or spreadsheets for example. The Capability Database can be accessed at the same workstation terminal used for further analysis by finite element software, or further geometry definition using CAD software and is in effect an on-screen technical guide for these processes.

The capability acquisition process is one of continuous improvement and as progress is made the database can be updated or expanded to suit. The Capability Database facilitates concurrent engineering by being accessible throughout the definition and verification phases of the shaft development life cycle, and the capability acquisition activities are in themselves concurrent with the mainstream project activities.

3.4 Experience Database

This is conceptually very similar to the capability database except for the contents. The Experience Database is populated with a catalogue of experience to facilitate the application of the relevant methods and capability, and also contains prompts for identifying and minimising risk in the process. The Experience Database is a facility that is accessible throughout the concurrent definition, verification and manufacture phases. It encourages a right first time end product, which improves quality and minimises the need for changes late in the development cycle.

3.5 Single Electronic Definition

The concept of the Shaft Key System is based on all processes having access to the same standard of electronic geometry definition of the component for better quality assurance and concurrency. Traditionally CAD models of shafts have been created explicitly by sweeping 2D sections around the shaft centre line and then adding or subtracting other features to complete the final solid definition, as depicted in Figure 2. This method of modelling can be restrictive or time consuming when changes are required by not taking full advantage of the more versatile 3D modelling tools.

The recent availability of parametric modelling capability has been used to good effect on shafts to improve the process. The model is created generally as before but constructed with associativity between the geometric entities to generate an implicit or parametric model of the component. The benefit derived from this approach is that the model can be modified by altering one or more parameters and the model is then regenerated automatically by the software according to the associativity stored in the model. The analysis models, detail definitions and manufacturing definitions are also linked to the central model to enable them to be created and updated efficiently. The system allows the component model and associated models to be worked on concurrently. The modelling history becomes divided into sub-sections that are more easily understood and the full range of solid modelling tools is available to slicken the modelling process. The philosophy is that all modelling and creation of detail and manufacture definitions is the responsibility of one modeller who will respond to the requirements of the various internal customers making use of the geometry and specialist analytical models. This can be achieved for simple components , but for a typical engine shaft geometry it is beyond the capability of current modelling techniques to produce a definitive component solid model typical of that in Figure 3, with a full suite of detail and analysis solid models as a truly concurrent activity by a single modeller in a multi-user environment. Further developments in the application of parametric modelling will be required to enable the process to be operated as intended. For a real shaft the current process is to create a definitive component model parametrically and then to create the detail and analysis parametric models with associativity to the definitive model.

In order to maximise the benefits to users accessing the electronic model the parametric relationships built in to the model need to be selected to suit the requirements of the various applications using the data. It is advantageous to minimise the dependency within the model by reducing the number of references between geometric entities created in the parametric history of the model. This results in a more robust model which can have entities removed with minimum disruption.

In the Shaft Key System the use of parametric modelling facilitates concurrent engineering by ensuring every part of the process has simultaneous access to the same standard of shaft definition with the required relationships built in to enable each application to generate the analysis model suitable for that process, thus ensuring that the added value is consistent at all stages in the component development life cycle. Consequently there will be closer links between the engineering and manufacturing functions and improved concurrent engineering resulting in a better match between the end product and the initial intent.

Parametric modelling within the Shaft Key System will enable better and faster evaluation of alternative design proposals by all functions simultaneously, faster definition of special development engine shafts modified for instrumentation for example and also gives the opportunity for earlier definition of similar shafts for the next project.

3.6 Analysis Software

The thermal, stress and dynamics analyses are generally conducted using finite element computing software. The modelling requirements can be subtly different for each application and this needs to be considered when deciding the construction and parametric relationships in the single electronic definition that is central to the Shaft Key System. Some analysis packages require minor specific modifications to the geometry or may need the parametric model to be in explicit form prior to the finite element mesh being generated. These are intermediate steps that can be introduced with relative ease between reading the central parametric model and proceeding with the analysis. Modelling in this way enables increased concurrent activity during the analysis phase of a project resulting in improvements in lead time and ensures the analysis models all relate to a consistent standard of geometry definition. The stress and thermal analyses use the same finite element software albeit in different ways and the use of automeshing facilities enable more efficient use of engineering resource resulting in consistent added value earlier in the definition and verification phases.

3.7 Detail Definition

In a traditional process the detail definition (or set of drawings) starts after the initial definition and usually finishes near or after the completion of the analysis phase. With the Key System approach this activity can begin when the parametric definition is complete, in line with the start of the analysis phase. The benefit gained by using parametric modelling is that geometry modifications following the analysis can be incorporated in the detail definitions more easily and quickly than detailing an explicit definition. A shaft modelled as an assembly of parametric parts needs to be dimensioned in a different manner to an explicit model. Views of the component are created differently in the parametric environment to enable parametric links to be created between the model and the drawing sheets. Part sections shown for clarity of detail can be generated more easily than in explicit. The generation of dimensioned detail views or drawings becomes more of a simultaneous activity as a consequence of parametric modelling.

3.8 Manufacturing Stage Definitions

The manufacturing phase of the shaft development life cycle requires a suite of definitions or drawings, in addition to the normal set of details, to define datums and other temporary features to assist the machining processes at the intermediate stages within the overall manufacture process. With a parametric model, the intermediate definitions can be generated with a degree of automation depending on the level of similarity between the part-machined shaft and the fully finished component. Again, for this activity the many benefits of reduced time scales can be realised by making it more concurrent than the traditional process.

4.0 CONCLUSIONS

The Shaft Key System is an integration of technology and software adapted from a legacy of disparate systems, improved and combined with new facilities to encourage concurrent engineering throughout the definition, verification and manufacture of the component resulting in a shorter time to market with a similar level of resource, reduced costs and a right first time end product. The benefits to the business are achieved in several ways.

Electronic data is more easily and quickly distributed and managed than paper documents. Revisions can be effected simultaneously and consistent data can be accessed throughout the company network resulting in improved quality.

Parametric modelling enables full use of 3D modelling capability and faster production of the dimensioned detail definitions required by the engineering and manufacturing processes. Similar alternative designs can be evaluated more efficiently with parametric modelling. Similar shaft definitions for the next project and special development engine standards can also be generated more easily.

Using the same source of electronic data throughout the component development life cycle achieves improved cross-functional involvement by enabling a consistent set of process models to be used. Consequently there is an increase in added value earlier in the process and better incorporation of manufacture and assembly requirements at the early design phase.
The increase in synergy between functional groups results in reduced cost and lead time, improved quality and faster response to customer requirements.

For continuous improvement there needs to be the same priority attached to capability acquisition as project realisation. The increase in concurrent engineering releases more time and resource to acquire capability and improve the Key System methods and databases.

Acknowledgements

The authors wish to thank Rolls-Royce plc for permission to publish this paper.

5.0 REFERENCES

5.1 T Broughton , "Simultaneous Engineering In Aero Gas Turbine Design And Manufacture", (report PNR90890), 1992, Aerospace Group, Rolls-Royce plc, Derby, UK.

5.2 Daizhong Su , "Concurrent Design Of Power Transmission Systems", Drives and Controls 96 conference, Telford, UK, Kamtech Publishing Ltd., 1996.

S520/002

FIGURE 1 Overview Of Shaft Key System

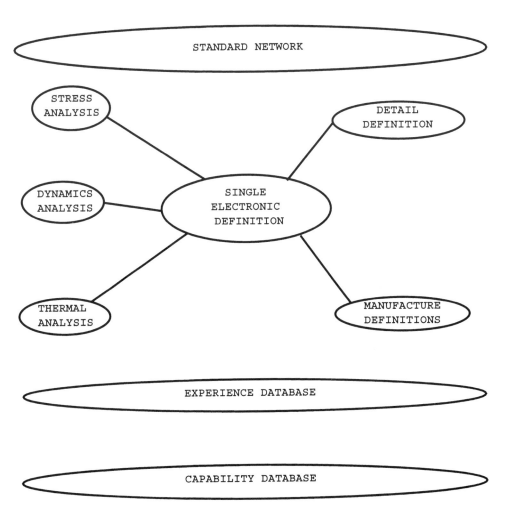

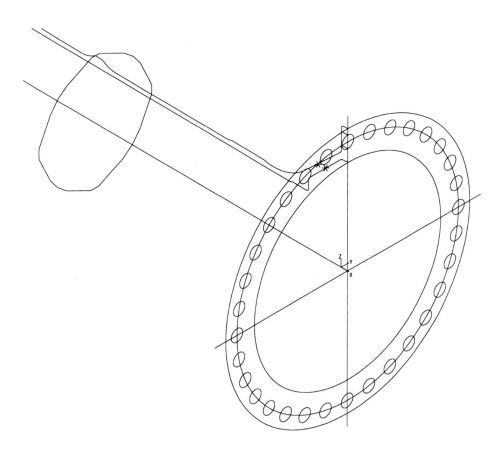

FIGURE 2   Typical Shaft Model Using 2D Sections In 3D Space

FIGURE 3   Typical Shaft Model Using Parametrics

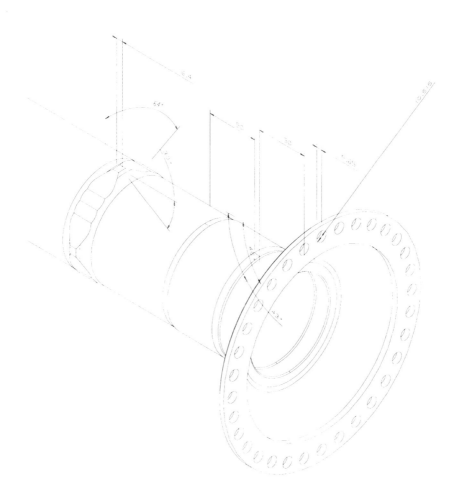

A new approach to transmission concurrent engineering

B JAMES, P POON and **L LACK**
Romax Technology Limited, Newark, UK

1. SYNOPSIS

Over the past ten years, much has been written about external pressures on a company, demanding reduced time-to-market, increased quality and reduced cost. The idea of Concurrent Engineering was created in order to solve many of these problems, and a host of Computer Aided Engineering (CAE) applications has been developed in order to implement the concept.

This paper shows that the natural progression is towards software that is more-and-more dedicated to specific tasks. The software package, RomaxDesigner, is a logical progression of this process and the advantages of such an approach over the current batch of general CAE tools are discussed.

2. CURRENT PROBLEMS FACING ENGINEERING DESIGN

The increasingly tough demands on Engineering Design are well understood. The life cycle of products is reducing as the global market becomes a reality, and the customer demands higher levels of quality along with lower costs.

A customer will expect to receive many quotations before placing an order and each candidate design must be validated against the intended application. A technical quotation of high quality and detail reflects well on the candidate and will fill the customer with confidence. The creation of these many quotations increases the workload on the design process.

As competition gets keener, a successful product needs to be optimised, both technically and commercially, for a particular application or niche. Hence, a company has to develop products for each application. Again, this increases the demands on the design process.

The concept of Concurrent Engineering was developed to improve the performance of the design process by addressing the issue of time-to-market for new product introduction.

However, it has been shown that the generation of technical quotations and the validation of design variants are just as important. The following section describes how Computer Aided Engineering has developed and discusses how the different approaches achieve these tasks.

3. THE USE OF COMPUTERS IN ENGINEERING

3.1 The Evolution of CAE tools

The first analysis routines were written in computer code. FORTRAN was an early tool and is still good for creating mathematical models of systems. Next came various mathematical packages, which made creating mathematical models easier. Then came Finite Element Code, Kinematic Modellers etc., which are tailored towards the creation of specific types of mathematical models.

If you compare each generation with the one before you see the same changes:
- The tools become more specialised, becoming dedicated to a fewer number of specific tasks.
- This specialisation inevitably brings with it a reduction in the breadth of capability.
- However, it also means that the tool is easier to use and a user is more rapidly able to set up complex analyses.

To summarise, there is always a trade-off between breadth and depth of capability. Specialisation (increasing the depth of capability) brings speed- and ease-of-use for complex tasks. This more than offsets the loss of breadth of capability.

3.2 Why has this evolution occurred?

It can be seen that the forces pushing the move towards more specialised software are exactly those which drive Concurrent Engineering:
- Time to market – speed of response means a more dedicated tool is required.
- Quality - in an increasingly competitive marketplace the requirement for complex analysis is greater. If a tool is too generalised this becomes too difficult or takes time too long.

These two pressures are external to the organisation. However, a third pressure, which comes from within the organisation, is often ignored:

- Human Resources – in the climate of continual cost-cutting it is not possible for an organisation to enjoy the luxury of experts in every area. Hence, Engineers are becoming multi-skilled generalists, yet they are still required to carry out complex tasks.

In order to achieve the combination of speed and quality the Engineer needs tools that are as dedicated as possible to his particular task.

Thus, we have the three requirements for any CAE tool:

- Speed-of-use – designs must be created rapidly, and changes must be easy-to-make for successful design optimisation.
- Performance – the software must implement a high level of design analysis.
- Ease-of-use – non-specialist, multi-skilled engineers must be able to use the software without needing lengthy training or regular practice.

So how do current CAE tools meet the requirements of the multi-skilled engineer for concurrent engineering?

4. DIFFERENT APPROACHES TO CONCURRENT ENGINEERING OF TRANSMISSIONS.

4.1 The application of conventional CAE software

Conventional CAD and CAE packages are common in most engineering companies, and over time the integration of the two has brought the engineering disciplines of design and analysis closer together. This is clearly an essential step in the implementation of Concurrent Engineering.

However, such tools are still extremely generalised. Solid modellers can define any geometry; FE codes can be applied to any structural or thermal analysis. This capability is substantially broader than is required for the majority of transmission engineering. Yet at the same time this capability is too shallow.

An example of this is in the area of bearing analysis. Rolling element bearing models should include load sharing between elements, radial internal clearance, rib loads and any combination of axial and radial loads and over-turning moments. FE models of bearings are extremely difficult to set up, requiring substantial expertise in both FEA and bearing analysis. Consequently, many companies do not bother, with the inevitable result that their analyses are inaccurate.

Hence, we see that conventional CAE tools in some way fail to meet all three of the essential requirements. The problem is that their capability is still too broad.

As we have seen above, excessive breadth of capability brings with it excessive complexity. This makes it difficult for the multi-skilled engineer to use. Additionally, complex models can be time-consuming to set up.

Further evidence of this is the existence of specialist operators of solid modelling packages. These people are able to make a career as sub-contractors, driving CAD packages and commanding ever-higher rates as they move from company-to-company.

FE packages are not so complex, but are still too complex for draughtsmen to operate. Hence, it is normal to find a "division of labour" between design and analysis that runs in direct opposition to the basic principles of Concurrent Engineering.

Again, we see that conventional CAE does not serve the design process in the required manner since it fails to meet at least one of the requirements: ease-of-use by generalist engineers, rapid design creation and advanced engineering performance.

4.2 Development of a tailored system

It has been shown that the trend is for software that is increasingly dedicated to a specific task. The idea is to develop a tool that is tailored to carrying out the required tasks, thus achieving the goals of Concurrent Engineering. There are a number of options for how such a system could be constructed for Transmission Engineering.

4.2.1 In-house versus Commercial Systems

A bespoke, in-house system is costly to produce, demanding a complex software framework and intensive implementation. The software skill needed to construct such a system is rare and tends to be the province of specialist centres-of-excellence. The risks of developing and implementing even a small in-house system are high, with cost and time over-runs common.

Increasingly companies are looking to commercial software companies for help. They develop systems for niche market sectors and take advantage of economies-of-scale since the large development costs are borne across many clients. They also have the software and implementation expertise.

4.2.2 "Knowledge Based" or "Knowledge-plus-Analysis"

Some companies have attempted to automate all their existing engineering activities by implementing them into a Knowledge Based System (KBS). In an attempt to protect against the retirement of experts, their knowledge is distilled into a set of design rules that are made available to the engineer who is involved in a given task.

However, the creation of such a rule-based system can cause problems, since there will be disagreement between different experts within a company. If an expert's opinions are ignored or compromised, he will inevitably resist using the system when it is completed.

Even if it is completed and accepted, the reliance on existing rules and methods automatically causes stagnation within an organisation. Continuous improvement ceases and the development of new technologies and methods stops. The organisation is now constrained into producing a standard product while all its competitors continue to evolve.

There is one final, inevitable problem with any rule-based approach. Rules stand still, whereas analysis progresses. One major manufacturer developed a multi-million pound KBS for transmission design. However, they found that a key part of the system, dealing with gear tooth optimisation, was made obsolete overnight by a superior programme developed externally.

The new method, which selects optimum gear designs based on a genetic algorithm, is now the preferred method and the rule-based system has been discarded. The lesson is that such systems are only practicable where no improvement is possible.

The application of in-house knowledge needs to be accompanied by analysis, which frees the engineer from current methods and allows him to look at new possibilities. This approach, of using "Knowledge-plus-Analysis", allows a step-by-step approach to change. It is paced by the development of new ways of working, yet incorporates the company's legacy data as well as the adoption of new technology.

4.3 Introduction to RomaxDesigner

RomaxDesigner is a software tool that has been developed in an attempt to address all the problems above. It is focussed on the tasks of power transmission engineering and makes use of the fact that the majority of power transmission involves standard components – shaft, bearings, gears, clutches, belts etc.

A common approach can be applied to problems in the automotive, aerospace, industrial and robotic sectors, hence the market is able to support a niche product that is both focussed yet highly developed. Consider how it addresses the problems outlined above:

4.3.1 Speed of design creation and modification

The model of a Transmission is created, not using geometrical entities such as points, lines etc., but using engineering objects such as shafts, bearings and gears.

It is important to make the distinction between standard geometrical objects and engineering objects. Figure 1 compares the characteristics of how objects are defined in different CAE packages. In most packages the user is left to define almost every feature of the object, including geometry, physical properties, the relationship from one component to the next etc. All this takes time, and if these "boundary conditions" are not set or are incorrectly entered, the results will be erroneous.

In a RomaxDesigner object, such as a bearing, the object is selected from an on-line catalogue with all the geometry defined. The object is imbued with all the engineering properties that are relevant to the functions under consideration. The process takes a couple of seconds.

The shafts, bearings and gears are assembled into a gearbox, from where the powerflow through the transmission can be defined. This automatically defines the loadcases for the analysis of all the components in the system. When a change to either the design or the loadcases is made, the analysis model automatically changes.

4.3.2 High level of engineering analysis

Once the bearing is located on the shaft, it automatically understands the relationship between itself and the rest of the system. The effect of misalignment on bearing life is automatically calculated.

Intelligent bearing models allow the load-dependent stiffness of the bearing to be calculated, taking into account both the stiffness of the shaft and the stiffness of the casing. (Due to the complex shape of the casing this is inevitably derived from FEA. This is an example where the flexibility of FEA is required).

All this affects the deflections and stresses in the shaft. The effect of misalignment on bearing life is calculated, taking into account roller profiling. An example of the analysis can be seen in figures 2 and 3.

Figure 2 shows the contact patch for a radial ball bearing under combined axial and radial load. It shows that the contact ellipse has moved off the raceway, resulting in high stresses and poor lubrication. Figure 3 shows the maximum stress on the heaviest loaded roller of a cylindrical roller bearing, and its dependency on roller profile.

The mesh misalignment of gears is automatically calculated from the shaft and bearing deflections and is included in the rating methods.

4.3.3 Ease-of-use by multi-skilled engineers
The use of Object Oriented Modelling and the setting up of Advanced Analysis makes the system easy to use by multi-skilled engineers.

Engineering objects are much more intuitive than geometrical entities. Underlying the system is an in-built understanding of transmissions. Since the software is dedicated to power transmissions, it understands the way the components are likely to fit together and the user is naturally "guided" through model creation.

All the "boundary conditions", so important in any analysis, are set. This means an engineer with relatively limited experience finds himself carrying out complex analyses within a few hours of using the software.

4.3.4 Combination of Knowledge and Analysis
Much has been said about the ability of multi-skilled engineers to use the software. An essential point is that it <u>enhances</u> engineering experience, instead of <u>replacing</u> it.

Any organisation has significant applications experience, and this should be used in conjunction with analysis results to gain a greater understanding of the engineering system.

A company that relies on applications experience plus basic analysis will inevitably fall behind. If a component fails it could due to a host of different reasons. Unless suitable analysis is carried out, the company will not understand the factor(s) that caused the failure. There is an important difference between curing-the-symptoms and understanding-the-cause, and any eventual solution will be sub-optimal.

An example of this comes in gear analysis. A company may use failure analysis to derive a material S-N curve and various applications factors. However, if, for example, the analysis fails to accurately account of mesh misalignment, it will result in inaccurate data being derived.

In bringing the best possible analysis to bear on any design it gives the Engineer greater understanding of the problem and ensures the experience he gains will be as accurate as possible. The speed of use of RomaxDesigner has meant that many companies are able to re-analyse past designs in order to understand why some designs were robust while others were not.

4.3.5 Rapid design validation and creation of quotations
Data is presented in an informative manner and can be easily formatted into a comprehensive report.

Once a design variant has been validated against the required operating conditions, all the data associated with the analysis is available. It can be exported and rapidly formatted into a report.

There is an important distinction between data and information. Instead of quoting meaningless stress values, RomaxDesigner informs the user about the performance (e.g. life) of the component.

5. APPLICATION OF ROMAXDESIGNER TO AEROSPACE TRANSMISSIONS

5.1 Creation of the model
The shaft is created as a series of sections, including tapered inner and outer diameters. Bearings are selected from an on-line catalogue. Alternatively, the user can define special bearings by entering the internal details of the bearing in a Custom catalogue.

Gear Pairs are defined with all the geometry required for rating to ISO 6336. They are then located on shafts.

The shafts are then assembled into a gearbox, with relative or absolute positions defined in either Cartesian or Polar co-ordinates. A view of a complete transmission assembly can be seen in figure 4. There is no limit to the number of shafts and gears within a transmission.

5.2 Definition of the Loads
Any number of power in- and outputs can be defined in a transmission, and a loadcase is created by defining the appropriate speeds and powers/torques. RomaxDesigner checks that the inputs are kinematically compatible and calculates the forces on all the components.

The transmission "Duty Cycle" consists of different loading conditions for different time-periods. RomaxDesigner allows any number of loading conditions to be defined.

5.3 Running the Analysis
Loadcases can be run either individually or as part of the overall Duty Cycle.

The shaft deflections and stresses are calculated, taking into account the calculated stiffness of the bearings and the stiffness of the housing.

The standard ISO Bearing life is quoted. However, this makes various assumptions concerning the loading arrangement of the bearing such as zero internal clearance, negligible misalignment and ideal load distribution. RomaxDesigner calculates the actual effect of all of these and quotes an Adjusted life. It is even possible to consider the effect of different roller profiles.

The mesh misalignment of each gear mesh is calculated and this is included in the rating calculation.

5.4 Methods of Design Optimisation

If a new gearbox is required for an application, the Engineer must propose a fully-validated candidate design as quickly as possible. This section investigates different ways in which RomaxDesigner facilitates this. The constraints on any design modifications vary a lot. A number of different examples are given:

5.4.1 Modifying the Duty Conditions

The easiest route to a new application is to use an existing transmission. It takes no more than a few minutes to re-define the Duty Cycle and re-run the analysis. The Engineer can then see if the existing transmission is suitable or if modifications are required.

5.4.2 Modifying the Component Design

At the component level the Engineer is able to make changes, the effect of which are propagated through the system. Examples are given below:
- Modifying a gear alters the rating results but can also affect the shaft deflection and bearing life.
- Modifying shaft geometry changes the shaft deflection which also changes the gear and bearing misalignment and hence life.
- Moving a gear pair or a bearing changes the loading arrangement of the whole system.

At each stage, the effect of any changes can be instantly seen by re-running the analysis.

5.4.3 Modifying the Transmission Layout

If the Engineer has sufficient freedom to modify the transmission layout, this can be achieved without a problem. A shaft can be rotated round another shaft with the Centre Distance maintained by means of the Gearbox Layout.

5.4.4 Automatic Gear Optimisation

RomaxDesigner allows the Engineer to modify a gear pair by hand and re-rate the gear. Normally a gear will undergo many design iterations before a solution is selected.

However, a gear pair can take many different values of pressure angle, module, addendum modification coefficient, addendum, dedendum etc. Effectively this constitutes a very large "design space" that can consist of tens-of-thousands of different candidate designs. Manual optimisation only finds local, not global, optimums, and a factorial search would take far too long.

Romax Technology has developed a Gear Search Algorithm, which uses Artificial Intelligence and Fuzzy Logic to search for the optimum solution. The user sets the geometry constraints (in terms of the allowable values of number of teeth, pressure angle etc.) and the targets (pinion and wheel bending safety factor, wheel contact safety factor etc.). An example of the targets can be seen in figure 5. The algorithm now searches for the optimum solution.

Typically after around 10 minutes, the software provides the 50 best candidate solutions for inspection. Inevitably, the candidates have different advantages/disadvantages, and the user is able to view their relative merits by means of a 3D view.

An example of the important characteristics of the 50 best candidate designs can be seen can be seen in figure 6. The user can then browse through the designs and select the most appropriate. It can be seen that radically different designs provide very similar performance, an indication of how important it is to search for the global, not local, optimum.

So far, the software has been tested against the "optimum" designs of experienced designers from many different companies. In every case, it has succeeded in locating superior solutions.

6. FURTHER EXTENSIONS TO THIS APPROACH

6.1 The Single Product Model

This paper describes a truly innovative approach to the problem of trying to optimise the productivity of Transmissions Engineers. The result is a unique product that has found uses in many different companies around the world.

However, further developments are still possible. The process of automating the engineering processes can be extended to include other methods and analyses, such that the whole system becomes a "Single Product Model". The central definition of the transmission is used as a reference database with many different engineering processes accessing the relevant data. This should extend from Concept Design to Manufacturing, from System Level performance simulation to Component micro-geometry analysis.

This will have major advantages for the corporation. Studies have shown that a major percentage of an Engineer's time is spent searching for data. With this all available in a central database such work is eliminated. Effectively it allows an Engineer to return to engineering.

6.2 Fatigue simulation

Romax Technology is looking to extend the functionality of its Single Product Model by linking to world-leading software from other companies. nCode International are world leaders in fatigue simulation software and consultancy, and a Technical Partnership Agreement has been signed between the two companies.

One of the first consequences of this collaboration will be the development of a link between RomaxDesigner and FATIMAS, which will allow the detailed fatigue analysis of shafts for any loading condition.

6.3 Analysis of shock loads

Another of Romax Technology's Technical Partners is AVL List, from Austria.

RomaxDesigner can be seen as a system that has been developed specifically for defining transmission layouts carrying out static analyses. Meanwhile TYCON, developed by AVL, is a system that has been developed specifically for the dynamic analysis of gear trains.

Since both products deal with similar engineering systems but carry out different analyses, Romax Technology and AVL are looking to link the two software packages together to create a Gear Train Dynamic Simulation.

Upon starting up, an Aerospace Transmission can often experience shock loads that can be critical in limiting the life of components. With links to TYCON and FATIMAS, the Single

Product Model will be able to calculate the magnitude of the shock load throughout the system and predict its effect on Low Cycle Fatigue Life.

7. SUMMARY

The vast majority of time spent in Transmission Engineering is not in "Clean Sheet" design but in the creation and validation of numerous design variants, and the preparation of detailed technical quotations.

The concept of Concurrent Engineering was created to enable a corporation to compete in an increasingly competitive global marketplace. The requirements for rapid product development have meant that many engineering processes are carried out using CAE.

It has been shown that there are three principle requirements if any software package to successfully facilitate Concurrent Engineering:

- Speed-of-use – the task must be achievable rapidly.
- Performance – high goals are set and must be attained.
- Ease-of-use – non-specialists must be able to use the software with lengthy training or regular practice.

This inevitably leads to software that is progressively more dedicated to specific tasks.

For this reason it is no surprise that genuine implementation of Concurrent Engineering of Transmissions relies upon the use of software that has been developed to match the tasks of Transmission Engineers. This is the only way that the goals of Concurrent Engineering can be achieved.

RomaxDesigner is such a package. The use of Object Oriented modelling and Advanced Analysis means all three criteria are met. The result is an increase in the productivity of engineers that is far beyond that which is achievable by conventional CAE.

In the future, this approach is to be extended to cover the complete product engineering process, from Concept to Manufacture, from Overall System Layout to Component Micro-Geometry. This will yield further time savings as information will always be made available at the point of use.

FIGURES

- 3D CAD
  - Purely geometrical representation

- Solid Modeller
  - Spatial and material properties

- FE Analysis
  - Mechanical reaction to forces

- RomaxDesigner Object
  - Spatial and mechanical reactions
  - Functions - Load and Life; Speed, etc.
  - Relationships - Inner ring to shaft,
  Outer ring to housing, Misalignment, etc.

Figure 1. The bearing object as defined in RomaxDesigner

Figure 2. The calculation of the contact track in a radial ball bearing under combined axial and radial load

Roller profile	Stress Plot
None	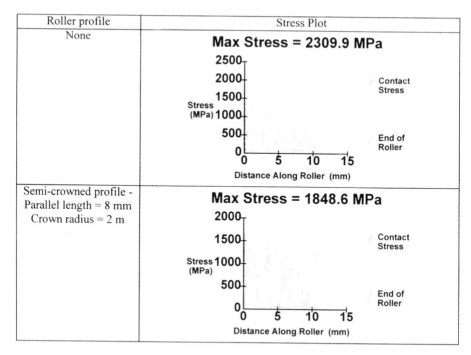
Semi-crowned profile - Parallel length = 8 mm Crown radius = 2 m	

Figure 3. The effect of roller profile on the maximum contact stress in a cylindrical roller bearing

S520/003

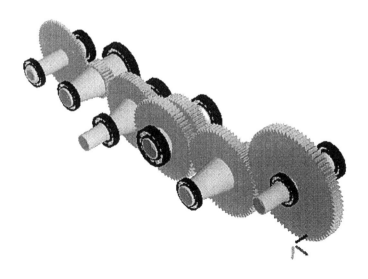

Figure 4. Three dimensional view of an aerospace transmission in RomaxDesigner

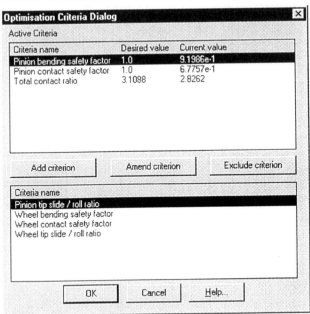

Figure 5. Defining the target parameters in the Automatic Gear Search Algorithm

Issues surrounding the synthesis of a software method for enhancing the design of involute splined shafts

A B POWELL
GKN Westland Aerospace, Somerset, UK

A software suite is being developed called SPADE, an acronym for Spline Predictive Analysis Design Evaluator. It is a prototype graphical user interface software tool whose objective is to check involute spline designs. It has an integrated database of existing designs. This allows easy review of designs on a comparative basis in a disciplined manner. The software takes account of the users experience. Instant Help and "tools" to guide users are provided throughout. It can be used on a PC with as little as 8 Mbytes of memory, under MS Windows 3.11 or Windows 95. A comprehensive design check can be made in as little as five minutes.

NOTATION

P   maximum compressive stress
V   maximum sliding velocity
HCI   Human Computer Interface
GUI   Graphical User Interface
LAN   Local Area Network

1   INTRODUCTION

Splined shafts are found as sliding fit connections in aerospace applications such as alternators, oil, fuel and hydraulic pumps. In helicopters, numerous splined couplings can be found. Main and Tail Rotor Gearboxes as well as other drive train assemblies, and the engine accessories mentioned earlier contain splined couplings.

2   THE DESIGN CHALLENGE

The challenge is to design splines that wear (fret) at low rates. Assembly tolerances introduce a misalignment between the splines at each end of the drive shaft. The effects of misalignment must be understood in order to synthesise a robust design which offers the most effective ratio of performance to weight. These effects provide contact conditions which reduce the load carrying capacity of the design. Contact between male and female splines change from a full surface at full alignment until eventually only two teeth take the load in any single rotation position. The contacts will be at the edge and corners of the spline teeth.

Since no material is perfectly rigid, elastic deformations result. These deformations of surface contact will occur at the contacting edges which decrease in load intensity as the distance from the initial edge contact increases. When the elastic contact is greater than the clearance between adjacent teeth, the load will also be carried but at a lower level than that of the initial driving tooth. A method of analysing these conditions is presented by Buckingham (1).

By examination, the maximum compressive stress (P) and the peak resultant axial sliding speed (V), we have the product PV that is an indicator of fretting susceptibility. An optimisation between the engaged spline length and these effects can be made, for any given misalignment. It is these calculations which have been embedded into SPADE, a software tool that can assist in the predictive analysis techniques required by designers in the competitive world class markets of today.

The calculation procedures in SPADE would be time consuming and error prone if performed by hand. So clearly, time and error potential for new designs can be minimised at design time. More importantly to an aircraft operator though, is the reduced likelihood of failure and aircraft downtime costs due to spline drives with higher than necessary wear rates.

3   A SOFTWARE METHOD

A software implementation of a mathematical method offers the strength of a guided approach to the analysis of a scenario without the analyst having necessarily to be an expert. One of the justifications behind producing SPADE, was to maintain a competence in spline design that may be used infrequently, whilst removing the need for costly training. The software attempts to provide a fast track route to any designer (whose competence and experience may vary vastly) the balanced design requirements necessary to create robust spline shaft design. In the main, the software ultimately is looking to reduce the risk of producing a poor design, via encodification of the Buckingham method for spline assessment.

4   SPECIFICATION

Many marketplace computer software packages are discipline independent. Word processors can be used by engineers for reports, legal secretaries for lawyer case notes or desktop publishing of a newsletter for a local scout group to name but three. The same can be said for spreadsheet applications, and even in the engineering world, finite element analysis (together with pre- and post-processing) can cover static strength, fluid flow, thermal transfer or even electromagnetic field analysis. To write an application specifically targeted at spline strength analysis is therefore not the norm.

Ideally, a team that would comprise of a mechanical engineer, a software database specialist, and an HCI designer would construct a specification. However, the project to date has been unable to draw the support that will allow the luxury of such a team.

By discussion and a literature review, it was concluded that to assist any designer to make the judgements necessary to effect robust designs, the calculation and review procedure would need to effectively address the following key features.

- offer assistance based on the designers experience
- minimise the number of inputs
- provide the user with feedback
- integrate a comparator database capability to store and judge design solutions
- require minimal training and documentation for the solution process
- utilise any beneficial HCI techniques to improve the delight of the user, ie make the program intuitive and rewarding

5 THE PROTOTYPE

Without the key supporting team players, it was found that the specification was evolving simultaneously with explorations into what could be encoded to result in a satisfactory GUI application.

The inability to secure support for this prototype demonstration application is probably an underlying factor in there being an apparent lack of GUI engineering applications being developed in companies. In the cases where they are, the code for analysis may be well executed, but the HCI is mediocre. It is this mediocrity that SPADE tries to overcome and in time through demonstration of the prototype attract support.

Figures 1 to 5 incl., show the principal screens with which the user would interface. Fig 1 illustrates an attempt to establish the likely user experience at the outset of the session together with the preferred unit system.

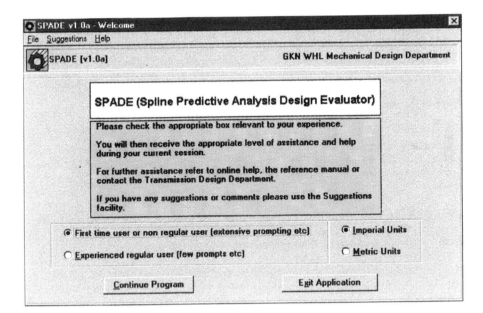

Figure 1 Introductory Screen to SPADE

The second screen, fig 2, shows a table of comparative values, the rightmost representing the limit of corporate experience. The left-hand values are given for two reasons. Initially they show typical values for the designer to use and secondly these can be used as default for quality assurance reasons.

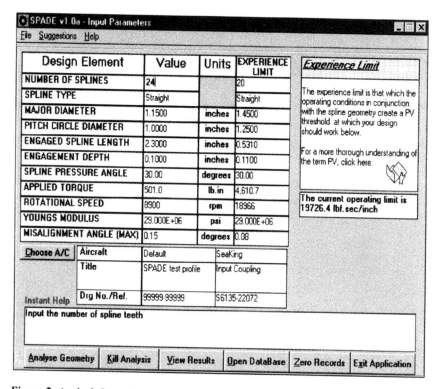

Figure 2 Analysis Input Screen

This will be particularly important when hardware platforms change, and as SPADE evolves within a continuous improvement cycle. An instant help screen is provided as an aide memoir to the designer at the foot of the window, particularly useful for infrequent or inexperienced users.

Action buttons are provided at the bottom of the screen, which when activated allow the designer to interrogate a database, perform an analysis or view results which may have already been obtained. The location of these button types is common to each screen.

Fig 3, over the page, shows a snapshot of the screen while the analysis engine is in process. Here it is important to provide the designer with feedback that the analysis is progressing, the rate of progress, and show that the analysis has not "hung". The method of analysis is to establish the sensitivity of PV vs. engaged length of spline form for a constant misalignment angle.

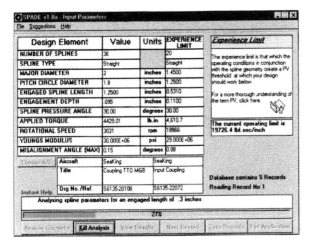

Figure 3 Analysis In Progress

Fig 4 shows how the designer is given a summary of the input values and how the sensitivity relates to the corporate experience. This is supported by some detailed stress and contact geometry information that experienced users may wish to peruse.

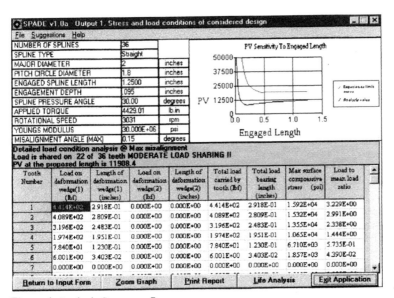

Figure 4 Analysis Summary Screen

For the inexperienced or uncertain user, a summary above the detailed spreadsheet of information is given. This information shows the load share and a maximum anticipated PV for the proposed length. A green background indicates an acceptable design and a red one unacceptable design. From the toolbar consistently positioned at the foot of the screen, an ability to zoom onto the graph is provided.

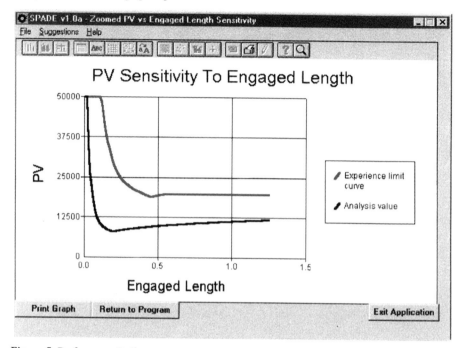

Figure 5 Performance Indicator

Fig 5 shows the "zoomed" screen. The experienced user can manipulate the graph from the iconised toolbar.

6 CONTINUOUS IMPROVEMENT CYCLE

A driving concept of SPADE is to offer both the user and the program itself, a continuous development process. To this end a suggestion facility has been provided, fig 6, which can be accessed from any point within SPADE.

Hopefully, this will drive SPADE development by a customer led requirements program. Although the prototype only sends a text file to a LAN directory, an E-mail function will be built in, to account for offsite and third party users.

At a personal level, the context sensitive help facility will provide a designer with an immediate corporate reference. A key benefit is that the unified database of central reference does not require circulation in the way in which manuals would, thus reducing an overhead cost.

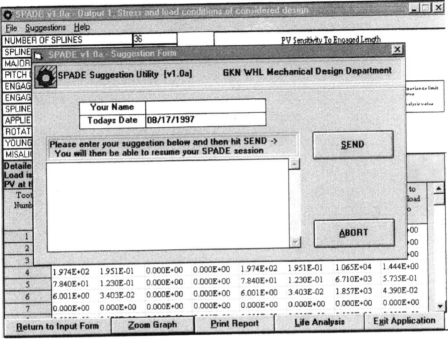

Figure 6 Integrated suggestion facility

7   USER TESTS

Initial user trials were conducted where the candidates selected had minimal or no experience in the design of spline shafts. The provision of the torque/speed/power facility, fig 7, was seen as a well conceived item. It saved finding a calculator and establishing what formula to use.

The trial also established that the use of windowed context sensitive help also improved the feel and user friendliness of the program.

A questionnaire also suggested that the software product would be remembered every time a spline was checked or designed. This in itself is a worthwhile benefit, since the frequency that a splined shaft is designed or checked is low. SPADE can therefore be expected to reduce the opportunity of mistake and oversight.

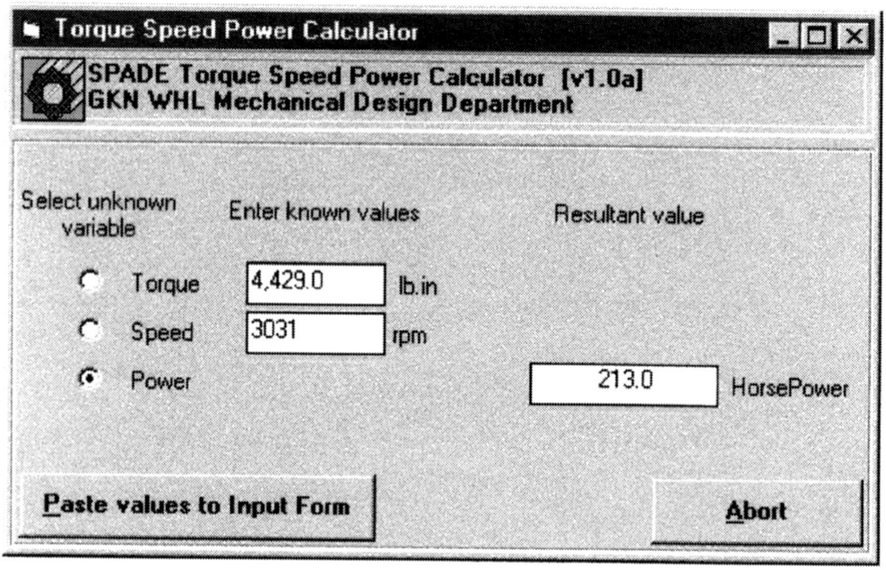

Figure 7 Design assistance tool

The trials established that it was possible to design a spline shaft from scratch in about 30 minutes. The interface was felt to have a key role in enthusing the user, the ability to see a design in a comparative light with key acceptance indicators forging a sense of satisfaction.

8 CONCLUSIONS

SPADE is a software tool with an inspiring interface. It is novel in that the number of knowledge based, comparative tools which consider the human user are few in the engineering design arena.

By careful development of the human interface and expansion of the number and type of analyses conducted, SPADE can achieve deskilling of the general design process achieved without compromise to the quality of the design.

This approach to HCI could further enhance many other prescriptive design approaches, particularly where greater levels of omnicompetent skills in each designer are required.

ACKNOWLEDGEMENTS

The author acknowledges the contribution of Mr L Crawley, and Mrs. J Crawley for programming advice, and Mr S Dare for contribution to testing. Acknowledgement is also given to Drs E Swain and I C Wright, Engineering Design Institute of Loughborough University for their advice.

REFERENCES

1   **Buckingham, E.** How to Evaluate the Effects of Spline Misalignments, *Machinery, August/September 1961*

Mathematical modelling of an aero gas turbine bearing chamber

K SIMMONS and **C EASTWICK**
Department of Mechanical Engineering, University of Nottingham, UK
S HIBBERD
Department of Theoretical Mechanics, University of Nottingham, UK

Synopsis
The computational modelling strategy described in this paper forms part of a project to create a mathematical model of a major part of an oil system, namely the HP-IP bearing chamber of a large three-shaft gas turbine. The chamber accommodates parts of two differentially-rotating, high-speed shafts and associated bearings, seals, oil jets, stationary components and scavenge system.

The complex two-phase flow phenomena within the bearing chamber are identified. An essential feature of the project strategy is the integration of experimental and computational work, which together will eventually lead to the development of a validated CFD model of the bearing chamber.

1. INTRODUCTION

This paper introduces a project to develop a verified CFD (Computational Fluid Dynamics) model of the HP-IP bearing chamber of a large three-shaft gas turbine. The commercial CFD code CFX (currently version 4.2) has been chosen and the project will require a number of man-years. The chamber is bounded by the High Pressure (HP) and Intermediate Pressure (IP) shafts of the engine and stationary components. Figure 1 shows a schematic representation of a bearing chamber. The HP shaft rotates typically at approximately 10 000 rpm and the IP shaft at 7000 rpm. The roller bearings, shown on Figure 1, require lubrication and cooling and this is provided in the form of an oil jet which is sprayed into the bearing chamber leading to under-race feed. The jet interacts with the core airflow breaking down into droplets and both are carried within a two-phase air/oil flow and deposited onto surfaces generating film flows. A complicating factor is that the chamber is sealed using labyrinth seals with a positive pressure such that there is a small air flow into the chamber. Excess oil (containing some air) is scavenged from the system at the bottom of the bearing chamber.

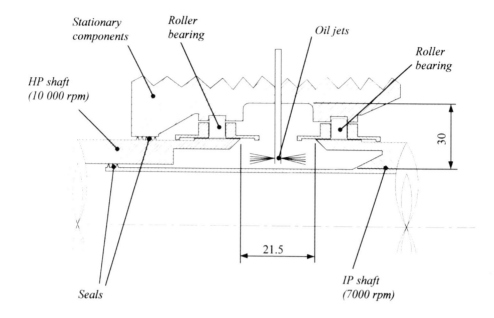

**Figure 1 – Schematic representation of HP-IP bearing chamber
(Generic, not to scale, dimensions in mm)**

It is extremely difficult to obtain reliable experimental information about the complex behaviour of the lubricant oil inside the bearing chamber. There are numerous problems that could arise if insufficient or oversupply of oil is provided to the bearings. Potential problems include carbon deposits in areas of poor lubrication, or even ignition of the air/oil mixture at points of high temperature. Taken to excess these problems could lead to bearing failure.

2. PHYSICAL PHENOMENA ASSOCIATED WITH BEARING CHAMBER

The flow within the bearing chamber is extremely complex. A number of physical phenomena have been identified , namely:

> oil jet behaviour;
> film modelling;
> film break-up and droplet entrainment;

The oil enters the chamber through two nozzles, which direct the oil into regions underlying the bearings. This oil will contain some air as a direct consequence of the scavenge and oil re-claim systems. Because of the high-speed rotation of the HP and IP shafts there will be considerable rotation of the air/oil mist in the bearing chamber. The induced airflow will probably deflect the oil jets and also may cause the jet to break up into droplets instead of remaining as a coherent jet. The behaviour of the jet is therefore one area of research. Experimental data will be used to provide understanding, which can be used to refine the CFD model.

As the oil moves around the bearing chamber (droplet deposition and film migration) films of differing thickness will form within the chamber. The thickness of such films will significantly affect heat transfer away from areas of high temperature. Film thickness and location will also be affected by the amount of airflow through the seals. Another area of research is therefore identified as the prediction and measurement of film thickness. Analytical methods for film thickness prediction will be developed and compared with experimental data. The combined calculation of core flow and thin films may be beyond the scope of current commercial CFD capabilities. It is the intention of this project to investigate the extent to which a commercial code can be used, and the areas for which new code is required.

It is suggested that the development and movement of droplets within the bearing chamber form a significant mechanism for the transfer of oil from inlet to scavenge. Film break-up and droplet entrainment are fruitful areas of research. In the bearing chamber there are many potential sources of droplet production. These include entrainment from films on rotating and stationary surfaces, droplets produced by jet break-up and droplets generated when larger droplets impinge on surface films. A CFD model of the bearing chamber would need to incorporate such droplet behaviour. Unfortunately it is not currently possible to predict droplet generation from a film or jet break-up using commercial CFD and in the current project analytical methods and experimental data will be used to supplement existing code capabilities.

3. OIL SYSTEM ISSUES

Once an adequate model of the flow within the bearing chamber has been obtained it will be possible to look at issues relating to oil systems. These include oil ignition, oil supply, carbon formation and oil sealing. Investigation into these complex areas will take place after initial model validity has been established.

Oil ignition could potentially occur in regions where the internal walls are sufficiently hot for ignition of the oil/air mixture to take place. It is thought that regions where the film is very thin, and regions where hot air enters the chamber through the seals will be most at risk. It is thought that the geometry of the bearing chamber will significantly affect the risk of oil ignition. A CFD model could be used to investigate the effect of geometry modifications on oil film thickness and oil/air ratios near seals.

The oil supply is also a clear area of interest. The design intention is to direct oil to the required areas in the most efficient manner. A CFD model would highlight areas of poor lubrication, and modifications to the oil delivery system could be analysed. A model incorporating full heat transfer capability could also highlight areas of poor cooling.

Carbon formation is obviously undesirable. Carbon deposits inhibit the transfer of heat away from hot surfaces and restrict the flow of oil around the bearing chamber. Carbon deposits are also indicative of temperatures higher than desired. A CFD model should be able to predict where carbon deposits are likely to form. Modifications to the geometry and oil supply can be tested on the CFD model to see whether there is improvement.

During normal operation, oil sealing is achieved by a small flow of hot air into the bearing chamber preventing oil flow out through the seals. The airflow is minimised to reduce the contamination of the oil with air bubbles. However, if the pressure across the seal is insufficient it is possible that oil will flow out through the seal. Also, if oil is allowed to build up near the seals then it may leak out. The CFD model will investigate pressure distribution within the chamber and also the likelihood of oil build-up in sensitive regions.

4. EXPERIMENTAL WORK

A CFD model can only be judged to be adequate if it has been validated against good experimental data. At each stage of the programme to develop a CFD model experimental validation will be sought. To this end a complementary experimental programme has been developed, and test rigs have been designed and are under construction. The main thrust of the experimental work centres on a custom-built test rig that reproduces the HP-IP bearing chamber. A number of existing and novel techniques will be used to measure the experimental variables.

Wall temperatures and pressures will be measured using existing technology. Pressure transducers and thermocouples will be installed at relevant locations within the bearing chamber. Oil film thickness will be measured using the capacitance technique. This is a relatively new technique that makes use of the different capacitance created by films of differing thickness. Capacitance transducers will be installed at relevant locations within the bearing chamber.

As previously mentioned, it is thought that droplet motion will be a significant factor in the movement of oil around the bearing chamber. Two complementary optical techniques will be used to obtain droplet size distributions at various locations and droplet velocities and trajectories. Particle Image Velocimetry techniques will be adapted for the extremely small bearing chamber using specially designed probes. Particle Image Velocimetry is a whole-field technique and will yield in-plane particle velocity information. A number of different planes within the bearing chamber will be investigated. A laser PDA (Phase Doppler Anemometry) system will be used in backscatter mode to obtain information about particle size distribution at various locations. As can be observed from Figure 1 the dimensions of the bearing chamber are small. The optical techniques, while not novel in themselves, must be adapted to the adverse conditions within the bearing chamber. Special probes have been designed and will be commissioned as part of the project.

5. COMPUTATIONAL STRATEGY

The basic modelling strategy consists of breaking the overall task into a series of smaller problems that can be investigated and verified at each stage. In this way confidence in the CFD model will build and the final model can be used to predict the outcome of design modifications without recourse to further expensive and time-consuming experimental work. A diagram illustrating the basic modelling approach is displayed in Figure 2. Initial CFD modelling will take place at ambient temperature and pressure. The added complexity of heat transfer characteristics will be added to the model later.

Stage 1: Single phase model of air flow

This initial modelling process models only the airflow in the bearing chamber and is an essential pre-cursor to future modelling work. Direct validation is not possible, as the bearing chamber will not operate without the essential oil flow and hence experimental data relating to single-phase airflow is not available. However, a simplified rig will provide data for a similar case, which will validate the technique.

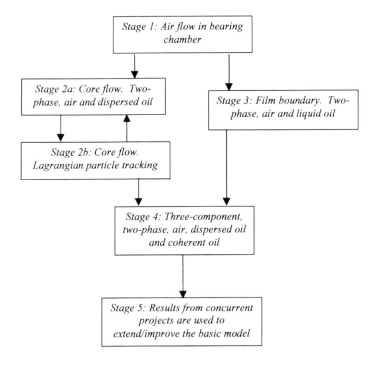

Figure 2 – Diagrammatic representation of solution process

Stage 2a: Two-phase core flow
The air and oil mixture within the bearing chamber (i.e. oil mist suspended in air, not including oil present as a film) is modelled using the Eulerian approach. In this approach transport equations are solved for both the air phase and the oil droplet phase. The solutions are coupled via interphase transfer co-efficients. Discrete droplet trajectories are not modelled, the two phases are treated as continuous. The oil flow field (i.e. the motion of the oil phase) can be compared with PIV data obtained experimentally. Oil-phase pressure data can also be compared with experimental data. When heat transfer effects are added to the model wall temperature distributions can be compared with experimental data.

At this stage of the modelling no attempt will be made to generate oil films on the bearing chamber surfaces. However, it will be possible to estimate film thickness and surface temperatures to incorporate into the model. Experimental data can be used to provide the necessary wall boundary conditions for this stage of the modelling.

Stage 2b: Lagrangian particle tracking
Using CFD software it is possible to track oil droplets through the computational model from the point of generation to the point of impact. This will generate information regarding where droplets land relative to size and location at creation. It is thought that oil droplets of differing size distribution will be generated from different locations within the bearing chamber. Droplet sources include jet breakdown, droplets entrained from films on HP and IP

shafts and stationary components, smaller droplets generated when larger droplets impinge on films and larger droplets created by the coalescence of smaller particles.

Stage 3: Film modelling

Stage 3 takes place in parallel with Stage 2. All surfaces within the bearing chamber will be coated with an oil film, the depth of film varying according to position. A study will be made to investigate the extent to which existing CFD is capable of modelling these thin films in isolation from the core flow. In the bearing chamber droplets will be entrained from the film, the droplet size depending on the film thickness and local velocity past the film. It will not be possible to actually model the entrainment process. However, in parallel with work utilising current commercial code, development work will take place to expand code capabilities.

Stage 4: Three-component, two-phase model

At this stage the expertise obtained in Stages 2 and 3 will be combined to create a model of the bearing chamber in which oil is present as both film and discrete droplets. This model will provide a state-of-the-art model of the bearing chamber utilising the full capabilities of current commercial CFD. It will then be possible to investigate the effect of oil and air temperatures on heat transfer and film thickness within the bearing chamber and compare data with experimental data.

At this stage it should be possible to use the CFD model as a design tool to investigate the effect of small changes to the oil delivery system, sealing arrangement, sealing air flow rates, scavenge flow rate etc.

Stage 5: Incorporate model extensions/improvements

Code development and model refinements will take place concurrent with the development of a CFD model utilising existing code capabilities. The results from these studies will be used to successively refine/improve the existing model. When sufficient agreement with experimental data is obtained the CFD model of the bearing chamber will become available as a full design tool. Expertise gained during the modelling process will permit models of other parts of the oil system to be modelled easily and with confidence. The duration of the design process will be considerably shortened and also the considerable expense of experimental development work will be substantially reduced.

6. CONCLUSION

A project is currently underway to develop a mathematical model of the HP-IP bearing chamber of a large three-shaft gas turbine, a major part of an oil system. The project will initially concentrate on the current capabilities of an existing commercial CFD code, CFX4.2. Experimental work and development work will take place alongside the modelling process. Experimental data will be used to validate the mathematical model at all stages of model creation. Development projects will be used to enhance areas where commercial code capabilities are poor, and add code where physical phenomena cannot adequately be modelled.

The project seeks to develop a validated CFD model of the HP-IP bearing chamber, sufficiently robust that future development work can be completed more quickly and at far less expense. Costly experimental work will be substantially reduced or eliminated in the case of small design modifications. The expertise gained during the project will be transferable to other components within the oil system providing a good design tool to complement current design methods.

S520/008/98

Three-dimensional technology in design and prototype manufacturing

K ZIMMERMANN and **F ADAM-RINGELSBACHER**
ZF Luftfahrttechnik, Friedrichshafen, Germany

SYNOPSIS

At ZF Luftfahrttechnik two methods of rapid prototyping have been tested. First to reduce time by creating 3D models and using this models for manufacturing of prototype parts. Second to use detailed Rapid Prototype-models, created by STEREOLITHOGRAPHY, directly for an investment casting process.

A significant saving in time was achieved with the first method. Maximum in time only can be saved if design work starts in 3D from the beginning.

With the second production-time for prototypes could be shortened considerably. For good results the co-operation between design, stress analysis and foundry must begin at an early step of design process.

1   INTRODUCTION

In the design process of technical equipment there is a tendency to reduce the time for development. At ZF Luftfahrttechnik (ZFL) two principal possibilities have been tested to achieve this target. The first is to reduce the time between development and manufacturing of prototypes. This is done by creating 3D models and using this models for manufacturing of parts for the prototypes.

The second is preparing detailed Rapid Prototype-models (RP-models) of parts with a complex geometry by using STEREOLITHOGRAPHY. This RP-models can be used directly for an investment casting process.

First the experience of ZFL with manufacturing of housing parts based on 3D files and after that the experience with STEREOLITHOGRAPHY will be introduced in this presentation.

2   3D-MODELS FOR DESIGN AND PROTOTYPE MANUFACTURING

2.1   Right Angle Propeller Gearbox for „Zeppelin NT"

In 1994 ZFL was requested to prepare a prototype propeller gearbox for investigation of the drive concept for the new „Zeppelin NT" which was under development at that time.

Picture 1: New Zeppelin „NT"
(source: Zeppelin Luftschifftechnik GmbH Friedrichshafen)

This gearbox had to be delivered to the customer within a very short time. Therefore ZFL decided to reduce the time for development by preparing 3D-models of the housing parts of this gearbox. Based on this 3D-models full working parts had to be manufactured from solid Aluminium material. Since the devices for manufacturing directly with 3D files are not available at ZFL the machining process had to be done by an external supplier.

2.1.1 Modelling

The 3D model has been developed on IDEAS Master Series 1.2 because this programme was available at ZFL at that time. Based on 2D CADAM drawings the geometry has been transferred manually to the 3D system and a 3D volume-model has been prepared. Problems occurred during modelling of radii and curves of the housing. Since inside radii will be produced by the tool geometry it has been agreed with the manufacturer that only outside radii will be modelled. Pictures of the FE-Model of the housing are shown below.

For checking of the parts and for a potential required final machining of the housing 2D drawings were needed. It however was not possible to derive these drawings from the 3D-model. All sections, prepared with the IDEAS „Drafting Layout Module" did not meet the requirements of the ZF standards. Thus these 2D drawings had to be prepared manually.

Picture 2: 3D-model of the Housing of the Zeppelin Propeller Gearbox

2.1.2 Data transfer and manufacturing

The manufacturer was working with the NC-System „MCAD". On a first step the Data transfer has been done with the IGES-interface (Version 5.0) in standard configuration of IDEAS. These Data however could not be used by the manufacturer. The surfaces were not transmitted as „trimmed surfaces". Since the manufacturer communicated that his NC-System did not work proper with IGES-Data no further attempt for transmitting has been done by ZFL. A new Data transfer has been done with a VDAFS interface (Version 2.0). The Data transmitted with this interface were applicable for the manufacturer.

With this Data one set of the housing parts of the propeller gearbox has been manufactured from a solid aluminium block. All short-term available solid Al-material had less strength than Aluminium alloy (A357). Due to the fact that the design of the housing was based on strength of A357-material this prototype gearbox has been released only for use on test rig. After manufacturing the housing on this way fits and surfaces with tight tolerance requirements had to be finally machined.

Picture 3: Housing of the Zeppelin Propeller Gearbox

2.1.3 Conclusion

After initial problems with Data transfer single housing parts of the right angle propeller gearbox for the Zeppelin NT have been designed in 3D and manufactured with the Data-set of the 3D drawing model. Fits and surfaces with tight tolerances had to be finally machined. Nevertheless it was possible to develop and manufacture the housing of the gear box within 4 weeks, from the first beginning until the parts were ready for installation.

This kind of rapid prototyping however was very expensive. Due to the fact that only aluminium-material with lower strength was available for manufacturing the gearbox could not be released for flight equipment. Therefore further parts, required for flight test equipment, have been made by casting processes.

2.2 Variable Torque Limiter

In 1995 ZFL started the development of a prototype of a Variable Torque Limiter for Actuation Systems. The developed Torque Limiter was designed for application in the FLAP Actuation System of large transport aircrafts and is combined with a Right Angle Gearbox.

Picture 4: Variable Torque Limiter

To check the progress in rapid prototyping with 3D technology ZFL decided to design and manufacture the housing of this Torque Limiter with this technology. The parts were manufactured from a solid aluminium block by an external supplier as well.

2.2.1 Modelling

In 1995 the ZF Concern decided to introduce „Pro/ENGINEER" as 3D system. Since ZF Luftfahrttechnik is a subsidiary company of ZF, Pro/ENGINEER (Pro/E) has been introduced as well.

Since the selected supplier only could produce the housing as rough machined part a 3D volume-model of the rough-housing has been created on Pro/E, Version 17, based on a 2D draft layout copy. This housing was very suitable for this task because of its geometry. There is a cylindrical section connected with a rectangular section. In the 2D drawing it was nearly impossible to create the transitional section. Since the 3D model was a volume model it did not cause any difficulties to create this transitional section. The main problem by preparing this 3D model was that the designer was not well trained and therefore not familiar enough with the Pro/E-system.

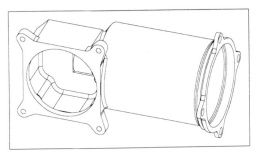 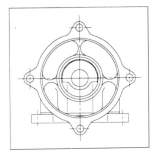

Picture 5: 3D-model of the Housing of the Variable Torque Limiter

The final machining of fit and surfaces with tight tolerance requirements had to be done at ZFL. For this task and for the final checking of the part 2D drawings were required. This 2D drawings could be derived directly from the 3D Pro/E -Data file without any problems.

Picture 6: Housing of the Torque Limiter

2.2.2 Data transfer and manufacturing

In this case the manufacturer was working with the NC-System „EUCLID". The Data have been converted with the IGES-interface and transferred to the manufacturer via the German telephone connection system ISDN. The transmitted Data were applicable for the manufacturer without any problems. Further action was necessary to work up the Data file to meet the requirements of the machining parameters regarding geometry.

With this Data-file all three pieces of the Torque Limiter housing, which were required for this development programme, have been manufactured from a solid Aluminium block. Determined by the above described geometry of the housing the parts were manufactured by a combination of drilling and cutting processes. After manufacturing the housing on this way not only fits and surfaces with tight tolerances had to be finally machined. Due to the small thickens of the flanges for the screw connection they were highly deflected. Thus this areas had to be machined finally as well.

2.2.3 Conclusion

It has appeared that parts with a complex geometry are easier to design in 3D than in 2D. There were approximately three working days required to prepare the 3D model. It took four weeks to manufacture the rough parts by the supplier and another week for final machining at ZFL. It however must be pointed out that in this case the target was not to get a prototype within the shortest time but to test whether a housing with such a complex geometry can be designed an manufactured directly in 3D technology. On this condition the target has been met.

This kind of rapid prototyping was very expensive in 1994. In 1996 it was not marked more expensive -for prototypes- than using a casting process.

At the presented examples a significant save in time for development and manufacturing of single parts was achieved. In both cases however the 3D model has been prepared on the basis of 2D drawings. For optimised process it is absolutely necessary that parts will be designed from the first beginning in 3D. Only by doing so a maximum time can be saved.

It must be understood that therefore a very good training of the design engineers in working with 3D technology is essential. A very important point is that the designer knows the 3D-System in details and that he is in the position to prepare the model in a way that the process sequences are tolerant against errors and for modifications. If the process sequence of the 3D model has not this tolerances a simple modification of a radius can lead to a complete breakdown of the whole model.

3   RAPID PROTOTYPING → THE QUICKEST WAY FROM THE IDEA TO THE FINAL PART

3.1   Hydraulic clutch parts for an APU gearbox

In 1994/95 ZFL was requested to develop and manufacture an APU gearbox for a helicopter prototype.

Picture 7: APU gearbox prototype

The APU gearbox has been designed for installation between the APU gasgenerator and the accessory gearbox (AGB - <u>A</u>ccessory <u>Gear</u>box) located in the main gearbox compartment of the Helicopter upper deck. The APU gearbox will rigidly link the APU gasgenerator to the AGB.

When the main engines of the helicoter are not operating and the APU is started, the APU output will accelerate and the hydraulic clutch in the APU gearbox will gradually engage. When the clutch is fully engaged, the APU gearbox will drive the accessories at the AGB, the externally attached fuel pump and the internal oil pump at the required speed. When the main engines are started, the drive from the main gearbox will accelerate to the speed of the APU, then the clutch in the APU gearbox will disengage allowing the APU to be shut down. The main gearbox will then drive the AGB and the accessories. With the clutch disengaged, the APU can be started independently of whether the main engines are running or not. After the clutch is fully engaged, the main engines can be stopped without any influence on the APU.

For the special requirements of the power transmission between the APU and the AGB and the high safety aspects in the flight conditions is the hydraulic clutch version the best solution. For the development of the hydraulic clutch parts, ZFL decided to split this workpackage in two phases.

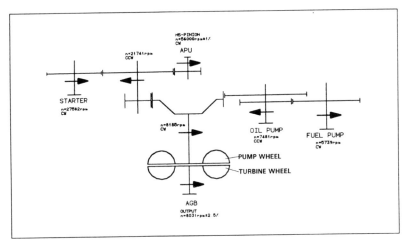

Picture 8: Geartrain of the APU Gearbox

In the first step it was necessary to fix and optimise the hydraulic clutch performance. The most difficulty was to find the best geometry of the pump- and turbine-wheel of the hydr.-clutch. For this optimisation process ZFL decided to manufacture the first prototypes of the hydr.-clutch with the Rapid Prototyping aluminium investment casting process. This first prototypes have been tested on the ZF research centre test bench.

In the second step ZFL manufactured the final clutch parts, with the geometry of step one, in the traditional aluminium investment casting process (material: aluminium alloy, A357) and has tested this hydr.-clutch in the original gearbox on the ZFL test bench.

3.1.1 Rapid prototyping for the investment casting

Rapid prototyping is a general expression for different methods including stereolithography, laser-sintering technology, LOM (laminated object manufacturing) methods and rapid tooling. It is possible to use all these methods for manufacture of lost patterns for investment casting production.

Rapid prototyping in investment casting offers special advantages to the customer in that the products can be introduced to the market within a very short time and development costs are minimal.

For the traditional production, a complete 2D-CAD drawing is normally drawn. The wax injection die, the so-called tool, is then made in accordance with this drawing. With these tools, the wax patterns are produced and then passed to the further traditional production process.

With the rapid prototyping process, the pattern is manufactured directly with the 3D-CAD datas after a preparation for the stereolithography machine. After the improvement of the pattern surface, the SL-model is handed over to the normal production process.

It is very clear to see the essential time saving for the manufacture of the prototypes with the rapid prototyping method.

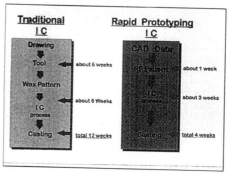

Picture 9: Time schedule for the traditional- and RP-investment casting process

(Source: Presentation folio from TITAL GmbH, Germany)

The most important advantages are the modification possibilities of the stereolithography model. Especially for the pre-series development and the development of very complicated cast parts, for which a test run must be completed before the final geometries can be fixed.

3.1.2 Principle manufacturing of the stereolithography model (SL model)

The XY-scanner first transfers the shape of the lowest slice to the surface of the liquid resin by a laser beam. The UV laser light hardens the synthetic material (liquid photo-polymer-resin) while this is happening. As the process continues, the working platform is immersed into the resin bath layer by layer. The process repeats itself continuously.

After the exposure of all layers, the working platform is raised and the finished model is removed.

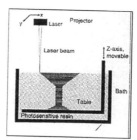

Picture 10: Principle of the stereolithography machine

(Source: Section of a prospect from Dornier GmbH, Germany)

3.1.3 Possibilities of the CAD data transfer

It is possible to transfer the CAD design data via a floppy disk, tape, CD-ROM or file transfer via modem.

Following CAD interfaces are possible: - IGES
                                          - VDA-FS
                                          - STL

3.1.4 Aluminium investment casting

In our example of the hydraulic clutch parts, ZFL prepare the 2D CADAM drawings of the pump wheel and the turbine wheel. A sub contractor of the casting supplier has been created the 3D model on base of our 2D CADAM drawings and the IGES files which we transferred via modem.
Based of this 3D CAD model it was possible to manufacture, within one week the first SL model (quick cast) for the further investment casting process.

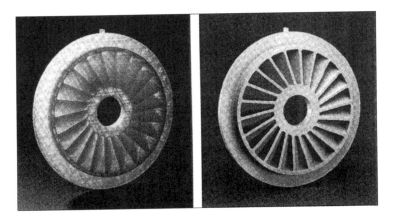

Picture 11: SL pattern and casted part of the turbine wheel

(Source   Section of a prospect from TITAL GmbH, Germany)

3.1.5 Conclusion

Out of the technical and economical view and using rapid prototyping processes for the investment casting, rapid prototyping can shorten production and development times considerably.
To do this, it is absolutely necessary that the co-operation between design, stress analysis and foundry begins at a very early step of the design process.

© With Author

S520/009/98

New technology for fast casing manufacture

N PEARCE
Development Group, Rolls-Royce Aerospace, Hucknall, UK
J FORFAR CEng, MIMechE
Advanced Engineering, Rolls-Royce Aerospace, Derby, UK

1.   SYNOPSIS

The investment cast Titanium intermediate casing is the major structural component in some large gas turbines, locating shafts, compressor, internal gearbox and radial drive for accessory power offtake.   It is longest leadtime component during the development programme of a new engine with the definition frozen very early in the cycle to enable tooling design and procurement. A new process being developed by Rolls-Royce, based around the build-up of material from weld beads in 3D space is described which could mean a substantial reduction in the lead time for the development of such casings. By eliminating casting tooling the design can evolve for a longer time while prototypes are made, and late change can be accommodated with ease. When a stable design has been achieved a "right first time" casing can be manufactured.

2.   BACKGROUND TO THE REQUIREMENT

In the mid-90's a Rolls-Royce internal time-compression study known as project Derwent idetified the intermediate casing of a large civil engine (Figure 1) as having the longest define/verify/make lead time of any component in a new engine programme.

Figure 1   The Trent 700 Intercase

Current large engine intercases are investment cast in titanium i.e. wax positive sectors are produced by injection moulding from large metal dies; these waxes are then sprayed with ceramic and the mould produced by melting out the wax. The long lead times associated with making the dies mean that information has to be given to the casting house a very long way in advance of the design being finalised, so effectively the design must be frozen before everything is known about the environment of the intercase and before all the design problems are resolved.

The casing is very complex and formerly took several months to draw, typically requiring 36 sheets of detail for the casting and another 30 for the machining. Stressing was also very labour-intensive, the casing sometimes not being stressed in detail until it was too late to correct problems cost-effectively.

What was required therefore was a way of applying the latest concurrent engineering techniques to the component, first finding a novel fast manufacturing method, and then setting up a key system revolving around essentially a single model of the component with all verification and manufacturing models derived efficiently from this.

3.   THE DEFINE AND VERIFY PROCESS
The problem with the define process is that the intercase is too large and complex for a solid model to be easily handled.   The approach adopted uses an adaptation of the CAMU (Concurrent Assembly Mock Up) system developed for the electronic mock-up of the external of the Trent, but with the difference that the various features (flanges, webs, vanes, holes) of a single component are modelled separately and viewed in together as an assembly. This means that not only can several modellers work on the same component at the same

time, but also that the model, consisting as it does of dozens of small parts, is not as cumbersome to manipulate as a single large model.

Duplicated features (e.g. an array of holes) can be stored as a single part and viewed simultaneously in many positions. This enables a very large and complex part to be defined on an ordinary CADDS workstation. The opportunity has also been taken to define all the parts parametrically, so that topologically similar casings can be defined quickly for derivative or developed engines by relatively painless scaling and local redefinition, rather than by starting over again.

Parts of the casing, or the whole, can be fused together to give sub-structures to suit the manufacturing process. These models can also be meshed to provide FE models of varying complexity; the ability to mesh across assemblies is also being pursued.

This brings us to the verify process, and again the problem is that the FE models produced are too large: as a comparison the Boeing 777 whole-aircraft model has 150 000 elements while the whole intercase FE model has 300 000. However, the larger the sector of the intercase that can be stressed in one go, the simpler the loads are to apply.

Accordingly our goal has been the ability to analyse the whole intercase overnight, processing the results on an ordinary workstation, and this objective is in sight. The current objective is to solve the thermo-mechanical analysis, using an in-house code, on a 4-processor parallel compute server in about 10 hours. Figure 2 shows a typical post-processor output for an earlier version of the model.

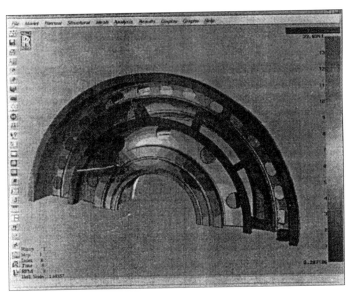

Figure 2   Half-intercase FE model solution

4.   MANUFACTURE

The new casing manufacturing process being developed by the manufacturing technology organisation could mean further substantial reductions in lead time. The Shaped Metal Deposition (SMD) development programme has been underway since early 1996, working on an idea from Rolls-Royce Bristol, and is funded by the Transmissions & Structures organisation.

4.1 Why is there a need for SMD?

Currently the design has to be frozen 18 months before the first casting is required, so that the complex and expensive casting tooling can be designed and manufactured. Machining is limited to points that interface with the rest of the engine and takes only 5 of the 18 months.

In response to the engineering need to reduce the overall lead time for the design and development of a new engine, Manufacturing Technology, having already conceived of SMD, proposed a development programme to take the process from the experimental stage to a production ready process. The aim of the programme is, therefore, to develop a method of manufacture, using the SMD process, for a Trent type intercase in Titanium 6Al 4V.

4.2 What is SMD?

SMD is the build-up of material from weld beads in 3-D space. The process starts on a baseplate and builds up into a component. The process is otherwise tool-less as the deposition process is planned to give a self supporting structure.

4.3 How will reductions in lead time be achieved?

SMD is currently being developed using a multi-axis robot (Figure 3) with a welding torch mounted in the arm. The robot is programmed off-line using CADCAM software and a robot simulation package.

It is these robot programmes that define the method of manufacture for the process and give the component its shape. It is currently envisaged that to improve the surface finish and to achieve walls thinner than the deposited beads, inter-deposition machining will be required. Therefore, the lead time for a SMD component is the time required to create the programmes and then deposit and machine the material. The time needed to design and manufacture the casting tooling has been eliminated. If SMD proves to be a viable production method, in addition to its role as a rapid prototyping tool, the entire casting tooling cost could also be saved.

SMD brings increased design flexibility as robot programmes can be altered just prior to deposition to accommodate late changes. This will ensure that the final geometry of the intercase will be optimised without the need for expensive changes to the casting tooling.

Figure 3   Multi-axis robot

5.   REAL SOLUTIONS TO REAL PROBLEMS

In August 1996, transmissions engineers approached the project to produce two steel components. These components were to form part of the interface between a Trent 700 module and the rest of the engine in the RB211-524G/H replacement, the 524GT. Due to the need to certify the engine as quickly as possible, the quoted 14 week forging lead time was not acceptable.

The first component (figure 4) was delivered for final machining within 5 weeks of this initial approach.   The first of the second type of component was produced from the CADDs geometry within one week - this included deposition, interstage machining and heat treatment. In total, three of each component were delivered over a 10 week period.

Two sets of these components were built into test bed engines and ran during the certification testing.   The engine was successfully certificated in 1997.   On strip of the engines, no problems were found with the components, so one set was built back into the 2000 cycle endurance engine.

Figure 4   The finished 524GT component

Following this success, another project has recently requested three components be made by SMD for a rig test. These were two 1100 mm diameter flanges and a 1100 mm diameter ring in steel. These components were delivered in 8 weeks as opposed to a quoted forging lead time of 30 weeks.

In between these requests for components, the project has progressed by manufacturing a number of features of the front section of the intercase. This has included the main section of the front case which is 1400 mm in diameter and 400 mm tall. The aim is to finish these feature trials and then to integrate them into a full component. Once the front case has been successfully manufactured the rear section, which is a much more complicated structure, will be attempted.

The potential uses of SMD are far wider than just for the intercase and because of this, an order for another SMD installation is imminent. The role of this new cell will be to explore these other applications while allowing the first cell to concentrate on the primary objective of the SMD programme, the intercase.

6.   CONCLUSION

The benefits of the improvements to the define/verify process are already being felt in current new engine programmes. The development of SMD as applied to large casings is ongoing: once a method of manufacture has been proved the full cross-functional benefit of a truly concurrent process will be achieved. Additionally, a decision can be taken as to whether the manufacturing process remains a tool for shortening the development lead time, or is economically viable for full production.

S520/009

S520/010/98

Best practice in managing change for concurrent engineering

S EVANS
CIM Institute, University of Cranfield, Bedford, UK

Synopsis

Many companies have improved their product development performance significantly. A number of these companies have been studied and common factors were sought. These factors were identified by the companies as key to the improvements they had achieved; the factors are primarily human and organisational in nature. The factors are presented in the form of a best practice chronology.

A method for increasing the likelihood that the factors are robustly and competently dealt with is also presented. The method is described as being suitable for most manufacturing companies seeking to make significant improvements to their product development process through multi-disciplinary teamworking.

1. GETTING BETTER?

Following decades of minor improvements in manufacturing as a process, industry has broken through the former performance barrier with new viewpoints such as Just In Time and Total Quality Management. Product development is now in the same transition phase as manufacturing was in the early 1980's. Some companies have been conducting their product development process successfully for many years - so is this change really different? The pioneers of superior product development processes have been joined by the fast-followers as the competitive advantages of faster time-to-market, of reduced launch costs, of higher reliability etc. have forced companies to learn quickly (1).

This paper is based on an analysis of some of those pioneers and fast-followers - companies who were selected on the basis of reported improvements in product development performance of over 25% in under 2 years. Many of the companies reported much higher improvements, as high as 400% improvement in launched product reliability for example. By visiting these companies, from many manufacturing sectors, and holding semi-structured interviews with the main personnel involved, we planned to learn:

1. what were the most important factors in achieving a significant improvement in product development process performance ?

2. Were the factors common to the successful companies, and did they form a clear pattern of implementation? And,

3. can the common factors, if they exist, be made usefully available to other product developers?

(Note that the emphasis on product development performance *as a process* is used to differentiate between successes based solely on technology or marketing. We wanted to discover the most important repeatable lessons, to help those who manage their own processes.)

The term Concurrent Engineering (CE) is used throughout to describe the methods the companies were using. They also used terms such as Simultaneous Engineering or Integrated Product Development to describe their own initiatives, vision or actions.

2. PEOPLE ARE THE PROBLEM...

The study has identified 10 main factors that contribute to successful implementation of CE. They are presented in a loose chronology to simplify explanation.

2.1 Preparing for the change
We observed all companies to undertake a planning phase when implementing CE. Many also used this phase to educate and prepare the rest of the organisation.

Factor one: Development and dissemination of management intention.
Senior management must develop a clear understanding of what CE entails. This was often done through visiting other CE practitioners and internal CE training. A prime concern raised internally and externally at this phase was how to manage the people through the implementation. A senior understanding of CE was disseminated alongside their intentions - how they planned to use CE. It was imperative that functional management (usually high, middle management) are actively involved in this phase (2). It was recognised by most organisations that the individuals who felt at most risk to the change *and* who were in a position to hinder the change were the highest level of functional management. The need to bring expertise (and hence, departments) together was identified early on as a major aim as well as a major issue. This was often presented as a 'team' solution with 'our culture' as the hindrance to team-working; both terms are commonly understood but sometimes used loosely. An important focus for this phase was the creation of a non-threatening environment in which people could talk openly about their concerns over CE.

Factor two: Change agents
Change agents were appointed by senior management to assist them in driving and monitoring the changes brought about by CE. They were often internal appointees in the form of CE Champions or CE steering committees. A significant proportion of their time was spent balancing the requirements of senior management, functional management and the

introduction of product development teams. Teams were seen as the prime way to reduce the barriers between 'functional silo's'. At this phase the emphasise on techniques and tools (such as CAE, QFD and FMEA) was reduced and the ability of any tool to support the teamworking process was emphasised.

Factor three: Pilot projects
Most companies implemented CE in a planned piecemeal manner. The initiation of one or sometimes two CE pilot projects, prior to expanding it to the entire organisation, was extremely common. Pilots were used to demonstrate success of the new way of working, as well as 'iron-out some wrinkles', and this helped gain senior management motivation to commit to similar changes on a wider scale. Most companies selected as pilot projects were chosen as projects with a low risk of failure to external reasons. They were typically given a high internal profile. Pilot projects were considered to be a good way of learning from real experience.

2.2 Creating an integrated product development environment
The change from a planning and preparation phase into the doing phase was significant. In particular, emphasis was given to creating the right 'environment' so that individuals with no previous experience of teamworking (and often the opposite!) could quickly begin to work differently.

Factor four: Integrated product development teams
As up to 80% of a products future costs can be determined during the early design stage (3), it is important that the disciplines which understand the impacts of early decisions are involved at this stage and throughout the product development process (4). The use of integrated product development teams, also known as CE teams, were widespread.

The selection of team members was generally done through negotiations between functional management and the appointed team leaders. This was often an informal process which considered mainly the technical ability and availability of potential team members. Little consideration was given to teamworking 'ability'. Similarly the use of early training in CE was not important; though many teams asked for, and got, training within 8 months of the project start. This seemed to based on the theory that getting on with it provided the most important learning and a better understanding of any training or selection requirements would emerge from the pilot. Some companies have since set up training and assessment centres.

In all teams there were both full-time and part-time members. Team size varied from single figures to thousands, obviously with varying degrees of structural complexity, though all used the team as a particular size of unit (e.g. large teams of hundreds made of working teams of 5 to 20). Most companies used the 'core team concept' where the main functions were represented by full-time members. These would typically be dominated by the design and manufacturing functions. This core team worked on the project full-time until it moved into volume production.

As well as having excellent project management skills the leadership role demanded sophisticated levels of people management skills. Some team leaders believed their technical skills should be treated as secondary. Most team leaders stated that their role should primarily be to manage and facilitate the team along its chosen path, occasionally playing a role as a technical team member. However this belief was not well practised and senior management continued to place greater emphasis on technical skills of team leaders during selection. Most team leaders came from the design function.

Factor five: Collocation
Sharing partial information incrementally is important to product development, where there are many task dependencies and uncertainty is high. If the team is collocated then informal communication can be improved dramatically (5). Most companies used a collocated core team; though some were constrained by geography with design and manufacturing sites being hundreds of miles apart. Collocation was also deliberately used as a signal that something different was happening, with identifiable 'team areas' which increased members identifying with the team.

Factor six: Team confidence
Senior management worked hard to ensure the teams were clear and confident about their roles and responsibilities; even negotiating them with the team and functional management. Unfortunately this work was often undocumented or formal, even though significant efforts had been put into it; thus leaving the various individuals re-clarifying roles with the CE champion when confusion appeared. It was clear that management needed to devolve sufficient resources and authority to give the team members confidence that they could achieve the project goals.

If the CE team felt supported they also felt greater ownership for their project tasks and late or poor quality delivery of tasks fell. This was especially important in the earliest design stages when the product specification was being agreed and the plan generated. Though apparently in an excellent position to support these tasks, and then to gain ownership of their contents, teams were rarely used here. Teams were typically formed after the specification and plan were agreed. Nevertheless many team members felt that they now have more discretion over defining their lower level plans than before.

Factor seven: Develop team environment interfaces
Having kicked off a new project with a multi-discipline team, the challenge moved to how to link this apparently 'one-off' project into the rest of the organisation. This was actively managed in order to smooth the delivery of the identified project by identifying the roles and responsibilities of the team with the rest of the organisation. In one example a test facility was not used to being given plans that were on time and the team soon learnt that facilities were not available when agreed!

2.3 Sustaining Concurrent Engineering
Within 6 to 12 months of the pilot project start the value of CE was confidently identified and agreed. Most companies then began programmes to accelerate the introduction of CE into the organisation as the normal method for product development. The companies experience varied greatly here, though most were tackling the factors identified.

Factor eight: Team based reward systems
Promotion, pay, measurement, reward and even educational systems needed to be reviewed to ensure they supported the new, team-based, way of working. Most organisations are still working on this and there is no clear consensus on the best way forward. Consensus on the difficulties raised for team members did exist.

Factor nine: Managing integrated product development projects

Most companies used a strong stage-gate process to ensure clear milestones and strong multi-disciplinary involvement in all stages. Most teams generated a strong desire to hit deadlines. Along with many other smaller differences, these changes needed to be reflected in the way the organisation and processes are structured. To date few companies have announced that they have achieved this and clear conclusions are not available.

Factor ten: Develop internal learning processes

Transferring the knowledge gained during pilot and early CE projects into the rest of the organisation is an on-going concern of many. With strong team identification it is reasonable to predict that functional specialists will find it difficult to hone their skills - because they see less projects - or to pass to their functional colleagues for use on other projects - through emphasis on team goals. Planned 'home' function time, workshops and Intranets are all solutions currently used.

Some companies followed the pilot projects with an internal review covering senior and middle management plus the product development team members. This large group commonly failed to organise the transfer of knowledge adequately, leaving it to the team members to disseminate the detail in their own ways

2.4 Integrating and ranking the factors

When analysed as a group of factors there were no single factors that appeared to be absolute. For each factor identified there was at least one notable success at implementing significant performance improvement without the factor. Neither was it possible to rank the factors for importance. However the overall pattern did correlate success with having the majority of the first seven factors in place.

This maps onto a common sense viewpoint that in the creative product development process people, their skills and their motivation are critical and that a good beginning will create a good end.

3. AND THE SOLUTION

The critical factors identified through research analysing the successful CE implementors have been assimilated into a tool called the FAST-CE implementation methodology. This has been used by number of companies and a workbook version of the methodology is available. Some of the ore interesting features of the FAST-CE implementation methodology are described below.

3.1 Action oriented

Many managers attend workshops to develop change management skills and ideas, but then go back to the same environments and relationships in the workplace. Training which has seemingly little relevance to the workplace is unlikely to achieve the desired changes in attitude and behaviours, as participants find it hard to transfer their learning. The implementation approach taken is based on the premise that learning is intimately related to action (6). The methodology therefore concentrates on actions that senior management must do to secure a successful launch of their CE implementation. These actions are delivered through a series of activities and exercises undertaken by senior management as a group. The exercises aim to help companies to create an appropriate infrastructure to manage and support a pilot CE project. After the pilot stage the workbook facilitates the expansion of the approach to subsequent product development projects.

3.2 Team based

Multi-discipline teams are the essential building blocks of Concurrent Engineering which bring together members from all relevant functions involved throughout the product development project. Members are selected to achieve the right skill mix. Any skills lacking are identified early and can be contracted into the team when they are needed.

The definition used for a team is *"a group that shares, and says that it shares, a common purpose, and recognises that it needs the efforts of every one of its members to achieve this".*

3.3 Learning through doing

A 1993 survey on UK Product Development (7) identified that even when senior managers were committed to CE, a major barrier to implementation was their persistent belief in a need for high levels of in-house CE expertise. This is contrasted to the same survey which revealed that the majority of CE expertise is created via a process of self-learning. For these reasons FAST-CE advocates a pilot introduction to CE, followed by a full project review which enables people to learn about CE by actually doing it. As individuals learn from their own experiences they will generate their own 'best practice solutions which satisfy their unique problems.

3.4 Gradual cultural change

The methodology enables a gradual cultural change in the organisation by a process which occurs in parallel to the actual development project. As individuals become familiar with CE, firstly through the workshops then, surrounded by others in the same state of development, through the actual project, new behaviours are tested. Within clearly stated boundaries these new ways of working are encouraged; later reviews will capture the natural decision making process which discards the poor behaviours.

3.5 The FAST-CE methodology

Based on one workshop for each phase identified in the ten factors, the methodology brings together all the relevant personnel into one room for two days each time. Initially focusing on senior management and their role in preparing the organisation for CE, issues such as pilot project selection, team membership, collocation, project selection and target setting are discussed and agreement gained. The clarification of roles and responsibilities is prominent as this most clearly communicates to team members the new expectations of the organisation.

In the second workshop both senior management, functional management and team members are present and it takes the form of a project launch. Emphasis is given to the new product with CE being advocated as a way of achieving the tough project targets. Activities involve all personnel in communicating and agreeing the specification, the project plan as well as time being spent on clarifying roles and responsibilities for all.

The final workshop is conducted when the pilot project is progressing with confidence. It has the same members and concentrates on the last three factors and on learning the lessons from the pilot.

4. CONCLUSIONS

After spending a great deal of time with excellent practitioners of CE, and analysing their comments, the absence of shocking or new insights is clear. Applying what appears to be common sense as taken even these organisations years or longer to evolve. But the common sense is still not common enough; this is probably because it is only common sense with hindsight. Most companies are faced with choosing where to take their product development processes, many will have heard of CE, but which of many alternative paths to take is not obvious.

All these companies will spend hours discussing people or culture, in the sure knowledge that they are key. But people and culture are massively complex issues, and the temptation to seek an equally complex solution is high. The experience of the leading CE practitioners is clear - look for the simple and the obvious. If you are having problems making departments talk then collocate them and give them a common target. Do not tell them to communicate, do not send them on communication courses, do not tell them how to communicate, just sit them next to each other!

REFERENCES

(1) Page A. L., *Assessing new product development practices and performance: Establishing crucial norms.* (Journal of Product Innovation Management. Vol 10. 273-290, 1993).

(2) Lettice, F.E., *Concurrent Engineering: a team-based approach to rapid implementation.* PhD Thesis. (The CIM Institute. Cranfield University, 1995).

(3) Andreasen, M.M and L. Hein., *Integrated product development.* Bedford: (IFS Publications Ltd.1987).

(4) Clark, K. B. and T. Fujimoto, *Product development performance - Strategy, organisation and management in the world auto industry,* Boston, Massachusetts: (Harvard Business School Press, 1991).

(5) Takeuchi, H. and I. Nonaka., *The new new product development game - stop running the relay race and take up rugby,* (Harvard Business Review, January 1986) pp.137 - 146.

(6) Revans, R. W., *Action Learning: New Techniques for Management.* (Blond and Briggs, London. 1980.)

(7) Costanzo, L., *Breaking out is hard to do.* Engineering, April 1993. pp. 25-26.

Acknowledgements This work was sponsored by the Engineering and Physical Sciences Research Council and was conducted by Dr Fiona Lettice and Dr Palie Smart.

Authors' Index